T0283606

Biofuels: Production, Assessment and Applications

Biofuels: Production, Assessment and Applications

Edited by
Mason Jones

Larsen & Keller
www.larsen-keller.com

Biofuels: Production, Assessment and Applications
Edited by Mason Jones
ISBN: 978-1-63549-044-2 (Hardback)

© 2017 Larsen & Keller

 Larsen & Keller

Published by Larsen and Keller Education,
5 Penn Plaza,
19th Floor,
New York, NY 10001, USA

Cataloging-in-Publication Data

Biofuels : production, assessment and applications / edited by Mason Jones.
        p. cm.
Includes bibliographical references and index.
ISBN 978-1-63549-044-2
1. Biomass energy. 2. Fuel. 3. Renewable energy sources. I. Jones, Mason.
TP339 .B56 2017
662.88--dc23

This book contains information obtained from authentic and highly regarded sources. All chapters are published with permission under the Creative Commons Attribution Share Alike License or equivalent. A wide variety of references are listed. Permissions and sources are indicated; for detailed attributions, please refer to the permissions page. Reasonable efforts have been made to publish reliable data and information, but the authors, editors and publisher cannot assume any responsibility for the vailidity of all materials or the consequences of their use.

Trademark Notice: All trademarks used herein are the property of their respective owners. The use of any trademark in this text does not vest in the author or publisher any trademark ownership rights in such trademarks, nor does the use of such trademarks imply any affiliation with or endorsement of this book by such owners.

The publisher's policy is to use permanent paper from mills that operate a sustainable forestry policy. Furthermore, the publisher ensures that the text paper and cover boards used have met acceptable environmental accreditation standards.

Printed and bound in the United States of America.

For more information regarding Larsen and Keller Education and its products, please visit the publisher's website www.larsen-keller.com

Table of Contents

Preface

This book explores all the important aspects of biofuels in the present day scenario. It discusses in detail the importance of biofuels and their different uses. Biofuels refer to the fuels which are produced by biological processes like anaerobic digestion and agriculture and not by geological processes. The different topics included in this extensive text cover the various forms of biofuel that are available such as biodiesel, bioethanol and biomass. While understanding the long-term perspectives of the topics, the book makes an effort in highlighting their impact as a modern tool for the growth of the discipline. Different approaches, evaluations and methodologies on biofuels have been included in it. This textbook, with its detailed analyses and data, will prove immensely beneficial to students involved in this area of biofuels at various levels.

A detailed account of the significant topics covered in this book is provided below:

Chapter 1- Biofuels are produced directly or indirectly from organic material, which include animal waste and plant materials. Out of all the total energy demanded by the world, 10% of the energy is provided by biofuels. The following chapter provides the reader with a basic understanding of biofuels.

Chapter 2- The types of biofuels discussed in this chapter are ethanol fuel, algae fuel, biodiesel and biogas. The most common biofuel worldwide is the ethanol fuel. While ethanol is particularly used in Brazil, one of the other types of biofuels, biodiesel is widely used in Europe. The topics discussed in the chapter are of great importance to broaden the existing knowledge on biofuels.

Chapter 3- Biofuels are seen as a means for replacing all of human energy needs from domestic fuel consumption to vehicular fuel. It can be the cause of reduction of greenhouse gases caused by the aircrafts and the aviation industry. The biofuel discussed in this section is sustainable aviation fuel. The focus of this chapter is the relevancy of biofuels in the aviation industry.

Chapter 4- Biomass can be used as a source of energy, and is mostly referred to plants or plant-based materials that are used for the manufacture of various sources of fuel. The conversion of biomass into biofuel can be achieved by different methods which are broadly classified in this chapter. Biofuels are best understood in confluence with the major topics listed in the following chapter.

Chapter 5- This chapter serves as a source to understand sustainable transport and the use of bioethanol. BioEthanol for Sustainable Transport (BEST) was a project that was introduced to support the usage of bioethanol as a vehicle fuel. The purpose of the use of bioethanol in transport is to have more flexible fuel vehicles and to decrease the amount of pollution caused by vehicles.

Chapter 6- This chapter provides a thorough understanding of the distinctive issues relevant to biofuels. Biofuels have several social and environmental concerns; some of the major challenges faced are the food vs. fuel issue, sustainable biofuel and deforestation. The content within this chapter attempts to summarize the disadvantages that should be considered in the use of biofuels, in order to encourage responsible consumption of the same.

Chapter 7- The production mechanism of biofuel is discussed in this chapter. Biofuel is derived from biomass and can be in solid, liquid or gaseous form. Waste management issues and pollution can be drastically reduced if we generate energy using biomass. The chapter strategically encompasses and incorporates the major components and key concepts of biofuel, providing a complete understanding.

I would like to make a special mention of my publisher who considered me worthy of this opportunity and also supported me throughout the process. I would also like to thank the editing team at the back-end who extended their help whenever required.

Editor

Introduction to Biofuels

Biofuels are produced directly or indirectly from organic material, which include animal waste and plant materials. Out of all the total energy demanded by the world, 10% of the energy is provided by biofuels. The following chapter provides the reader with a basic understanding of biofuels.

A biofuel is a fuel that is produced through contemporary biological processes, such as agriculture and anaerobic digestion, rather than a fuel produced by geological processes such as those involved in the formation of fossil fuels, such as coal and petroleum, from prehistoric biological matter. Biofuels can be derived directly from plants, or indirectly from agricultural, commercial, domestic, and/or industrial wastes. Renewable biofuels generally involve contemporary carbon fixation, such as those that occur in plants or microalgae through the process of photosynthesis. Other renewable biofuels are made through the use or conversion of biomass (referring to recently living organisms, most often referring to plants or plant-derived materials). This biomass can be converted to convenient energy-containing substances in three different ways: thermal conversion, chemical conversion, and biochemical conversion. This biomass conversion can result in fuel in solid, liquid, or gas form. This new biomass can also be used directly for biofuels.

A bus fueled by biodiesel

Bioethanol is an alcohol made by fermentation, mostly from carbohydrates produced in sugar or starch crops such as corn, sugarcane, or sweet sorghum. Cellulosic biomass, derived from non-food sources, such as trees and grasses, is also being developed as a feedstock for ethanol production. Ethanol can be used as a fuel for vehicles in its pure form, but it is usually used as a gasoline additive to increase octane and improve vehicle emissions. Bioethanol is widely used in the USA and in Brazil. Current plant design does not provide for converting the lignin portion of plant raw materials to fuel components by fermentation.

Information on pump regarding ethanol fuel blend up to 10%, California

Biodiesel can be used as a fuel for vehicles in its pure form, but it is usually used as a diesel additive to reduce levels of particulates, carbon monoxide, and hydrocarbons from diesel-powered vehicles. Biodiesel is produced from oils or fats using transesterification and is the most common biofuel in Europe.

In 2010, worldwide biofuel production reached 105 billion liters (28 billion gallons US), up 17% from 2009, and biofuels provided 2.7% of the world's fuels for road transport. Global ethanol fuel production reached 86 billion liters (23 billion gallons US) in 2010, with the United States and Brazil as the world's top producers, accounting together for 90% of global production. The world's largest biodiesel producer is the European Union, accounting for 53% of all biodiesel production in 2010. As of 2011, mandates for blending biofuels exist in 31 countries at the national level and in 29 states or provinces. The International Energy Agency has a goal for biofuels to meet more than a quarter of world demand for transportation fuels by 2050 to reduce dependence on petroleum and coal. The production of biofuels also led into a flourishing automotive industry, where by 2010, 79% of all cars produced in Brazil were made with a hybrid fuel system of bioethanol and gasoline.

There are various social, economic, environmental and technical issues relating to biofuels production and use, which have been debated in the popular media and scientific journals. These include: the effect of moderating oil prices, the "food vs fuel" debate, poverty reduction potential, carbon emissions levels, sustainable biofuel production, deforestation and soil erosion, loss of biodiversity, impact on water resources, rural social exclusion and injustice, shantytown migration, rural unskilled unemployment, and nitrous oxide (NO2) emissions.

Liquid Fuels for Transportation

Most transportation fuels are liquids, because vehicles usually require high energy density. This occurs naturally in liquids and solids. High energy density can also be provided by an internal combustion engine. These engines require clean-burning fuels. The fuels that are easiest to burn cleanly are typically liquids and gases. Thus, liquids meet the requirements of being both energy-dense and clean-burning. In addition, liquids (and gases) can be pumped, which means handling is easily mechanized, and thus less laborious.

First-generation Biofuels

"First-generation" or conventional biofuels are made from sugar, starch, or vegetable oil.

Ethanol

Neat ethanol on the left (A), gasoline on the right (G) at a filling station in Brazil

Biologically produced alcohols, most commonly ethanol, and less commonly propanol and butanol, are produced by the action of microorganisms and enzymes through the fermentation of sugars or starches (easiest), or cellulose (which is more difficult). Biobutanol (also called biogasoline) is often claimed to provide a direct replacement for gasoline, because it can be used directly in a gasoline engine.

Ethanol fuel is the most common biofuel worldwide, particularly in Brazil. Alcohol fuels are produced by fermentation of sugars derived from wheat, corn, sugar beets, sugar cane, molasses and any sugar or starch from which alcoholic beverages such as whiskey, can be made (such as potato and fruit waste, etc.). The ethanol production methods used are enzyme digestion (to release sugars from stored starches), fermentation of the sugars, distillation and drying. The distillation process requires significant energy input for heat (sometimes unsustainable natural gas fossil fuel, but cellulosic biomass such as bagasse, the waste left after sugar cane is pressed to extract its juice, is the most common fuel in Brazil, while pellets, wood chips and also waste heat are more common in Europe) Waste steam fuels ethanol factory- where waste heat from the factories also is used in the district heating grid.

U.S. President George W. Bush looks at sugar cane, a source of biofuel, with Brazilian President Luiz Inácio Lula da Silva during a tour on biofuel technology at Petrobras in São Paulo, Brazil, 9 March 2007.

Ethanol can be used in petrol engines as a replacement for gasoline; it can be mixed with gasoline to any percentage. Most existing car petrol engines can run on blends of up to 15% bioethanol with petroleum/gasoline. Ethanol has a smaller energy density than that of gasoline; this means it takes more fuel (volume and mass) to produce the same amount of work. An advantage of ethanol (CH_3CH_2OH) is that it has a higher octane rating than ethanol-free gasoline available at roadside gas stations, which allows an increase of an engine's compression ratio for increased thermal efficiency. In high-altitude (thin air) locations, some states mandate a mix of gasoline and ethanol as a winter oxidizer to reduce atmospheric pollution emissions.

Ethanol is also used to fuel bioethanol fireplaces. As they do not require a chimney and are "flueless", bioethanol fires are extremely useful for newly built homes and apartments without a flue. The downsides to these fireplaces is that their heat output is slightly less than electric heat or gas fires, and precautions must be taken to avoid carbon monoxide poisoning.

Corn-to-ethanol and other food stocks has led to the development of cellulosic ethanol. According to a joint research agenda conducted through the US Department of Energy, the fossil energy ratios (FER) for cellulosic ethanol, corn ethanol, and gasoline are 10.3, 1.36, and 0.81, respectively.

Ethanol has roughly one-third lower energy content per unit of volume compared to gasoline. This is partly counteracted by the better efficiency when using ethanol (in a long-term test of more than 2.1 million km, the BEST project found FFV vehicles to be 1-26 % more energy efficient than petrol cars The BEST project), but the volumetric consumption increases by approximately 30%, so more fuel stops are required.

With current subsidies, ethanol fuel is slightly cheaper per distance traveled in the United States.

Biodiesel

Biodiesel is the most common biofuel in Europe. It is produced from oils or fats using transesterification and is a liquid similar in composition to fossil/mineral diesel. Chemically, it consists mostly of fatty acid methyl (or ethyl) esters (FAMEs). Feedstocks for biodiesel include animal fats, vegetable oils, soy, rapeseed, jatropha, mahua, mustard, flax, sunflower, palm oil, hemp, field pennycress, *Pongamia pinnata* and algae. Pure biodiesel (B100) currently reduces emissions with up to 60% compared to diesel Second generation B100.

Biodiesel can be used in any diesel engine when mixed with mineral diesel. In some countries, manufacturers cover their diesel engines under warranty for B100 use, although Volkswagen of Germany, for example, asks drivers to check by telephone with the VW environmental services department before switching to B100. B100 may become more viscous at lower temperatures, depending on the feedstock used. In most cases, biodiesel is compatible with diesel engines from 1994 onwards, which use 'Viton' (by DuPont) synthetic rubber in their mechanical fuel injection systems. Note however, that no vehicles are certified for using neat biodiesel before 2014, as there was no emission control protocol available for biodiesel before this date.

Electronically controlled 'common rail' and 'unit injector' type systems from the late 1990s onwards may only use biodiesel blended with conventional diesel fuel. These engines have finely metered and atomized multiple-stage injection systems that are very sensitive to the viscosity of the fuel. Many current-generation diesel engines are made so that they can run on B100 without altering the engine itself, although this depends on the fuel rail design. Since biodiesel is an effective solvent and cleans residues deposited by mineral diesel, engine filters may need to be replaced more often, as the biofuel dissolves old deposits in the fuel tank and pipes. It also effectively cleans the engine combustion chamber of carbon deposits, helping to maintain efficiency. In many European countries, a 5% biodiesel blend is widely used and is available at thousands of gas stations. Biodiesel is also an oxygenated fuel, meaning it contains a reduced amount of carbon and higher hydrogen and oxygen content than fossil diesel. This improves the combustion of biodiesel and reduces the particulate emissions from unburnt carbon. However, using neat biodiesel may increase NOx-emissions Nylund.N-O & Koponen.K. 2013. Fuel and Technology Alternatives for Buses. Overall Energy Efficiency and Emission Performance. IEA Bioenergy Task 46. Possibly the new emission standards Euro VI/EPA 10 will lead to reduced NOx-levels also when using B100.

Biodiesel is also safe to handle and transport because it is non-toxic and biodegradable, and has a high flash point of about 300 °F (148 °C) compared to petroleum diesel fuel, which has a flash point of 125 °F (52 °C).

In the USA, more than 80% of commercial trucks and city buses run on diesel. The emerging US biodiesel market is estimated to have grown 200% from 2004 to 2005. "By the end of 2006 biodiesel production was estimated to increase fourfold [from 2004] to more than" 1 billion US gallons (3,800,000 m³).

In France, biodiesel is incorporated at a rate of 8% in the fuel used by all French diesel vehicles. Avril Group produces under the brand Diester, a fifth of 11 million tons of biodiesel consumed annually by the European Union. It is the leading European producer of biodiesel.

Other Bioalcohols

Methanol is currently produced from natural gas, a non-renewable fossil fuel. In the future it is hoped to be produced from biomass as biomethanol. This is technically feasible, but the production is currently being postponed for concerns of Jacob S. Gibbs and Brinsley Coleberd that the economic viability is still pending. The methanol economy is an alternative to the hydrogen economy, compared to today's hydrogen production from natural gas.

Butanol (C_4H_9OH) is formed by ABE fermentation (acetone, butanol, ethanol) and experimental modifications of the process show potentially high net energy gains with butanol as the only liquid product. Butanol will produce more energy and allegedly can be burned "straight" in existing gasoline engines (without modification to the engine or car), and is less corrosive and less water-soluble than ethanol, and could be distributed via existing infrastructures. DuPont and BP are working together to help develop butanol. *E. coli* strains have also been successfully engineered to produce butanol by modifying their amino acid metabolism.

Green Diesel

Green diesel is produced through hydrocracking biological oil feedstocks, such as vegetable oils and animal fats. Hydrocracking is a refinery method that uses elevated temperatures and pressure in the presence of a catalyst to break down larger molecules, such as those found in vegetable oils, into shorter hydrocarbon chains used in diesel engines. It may also be called renewable diesel, hydrotreated vegetable oil or hydrogen-derived renewable diesel. Green diesel has the same chemical properties as petroleum-based diesel. It does not require new engines, pipelines or infrastructure to distribute and use, but has not been produced at a cost that is competitive with petroleum. Gasoline versions are also being developed. Green diesel is being developed in Louisiana and Singapore by ConocoPhillips, Neste Oil, Valero, Dynamic Fuels, and Honeywell UOP as well as Preem in Gothenburg, Sweden, creating what is known as Evolution Diesel.

Biofuel Gasoline

In 2013 UK researchers developed a genetically modified strain of Escherichia coli (E.Coli), which could transform glucose into biofuel gasoline that does not need to be blended. Later in 2013 UCLA researchers engineered a new metabolic pathway to bypass glycolysis and increase the rate of conversion of sugars into biofuel, while KAIST researchers developed a strain capable of producing short-chain alkanes, free fatty acids, fatty esters and fatty alcohols through the fatty acyl (acyl carrier protein (ACP)) to fatty acid to fatty acyl-CoA pathway *in vivo*. It is believed that in the future it will be possible to "tweak" the genes to make gasoline from straw or animal manure.

Vegetable Oil

Straight unmodified edible vegetable oil is generally not used as fuel, but lower-quality oil can and has been used for this purpose. Used vegetable oil is increasingly being processed into biodiesel, or (more rarely) cleaned of water and particulates and used as a fuel.

Filtered waste vegetable oil

Walmart's truck fleet logs millions of miles each year, and the company planned to double the fleet's efficiency between 2005 and 2015. This truck is one of 15 based at Walmart's Buckeye, Arizona distribution center that was converted to run on a biofuel made from reclaimed cooking grease produced during food preparation at Walmart stores.

As with 100% biodiesel (B100), to ensure the fuel injectors atomize the vegetable oil in the correct pattern for efficient combustion, vegetable oil fuel must be heated to reduce its viscosity to that of diesel, either by electric coils or heat exchangers. This is easier in warm or temperate climates. MAN B&W Diesel, Wärtsilä, and Deutz AG, as well as a number of smaller companies, such as Elsbett, offer engines that are compatible with straight vegetable oil, without the need for after-market modifications.

Vegetable oil can also be used in many older diesel engines that do not use common rail or unit injection electronic diesel injection systems. Due to the design of the combustion chambers in indirect injection engines, these are the best engines for use with vegetable oil. This system allows the relatively larger oil molecules more time to burn. Some older engines, especially Mercedes, are driven experimentally by enthusiasts without any conversion, a handful of drivers have experienced limited success with earlier pre-"Pumpe Duse" VW TDI engines and other similar engines with direct injection. Several companies, such as Elsbett or Wolf, have developed professional conversion kits and successfully installed hundreds of them over the last decades.

Oils and fats can be hydrogenated to give a diesel substitute. The resulting product is a straight-chain hydrocarbon with a high cetane number, low in aromatics and sulfur and does not contain oxygen. Hydrogenated oils can be blended with diesel in all proportions. They have several advantages over biodiesel, including good performance at low temperatures, no storage stability problems and no susceptibility to microbial attack.

Bioethers

Bioethers (also referred to as fuel ethers or oxygenated fuels) are cost-effective compounds that act as octane rating enhancers."Bioethers are produced by the reaction of reactive iso-olefins, such as iso-butylene, with bioethanol." Bioethers are created by wheat or sugar beet. They also enhance engine performance, whilst significantly reducing engine wear and toxic exhaust emissions. Though bioethers are likely to replace petroethers in the UK, it is highly unlikely they will become a fuel in and of itself due to the low energy density. Greatly reducing the amount of ground-level ozone emissions, they contribute to air quality.

When it comes to transportation fuel there are six ether additives- 1. Dimethyl Ether (DME) 2. Diethyl Ether (DEE) 3. Methyl Teritiary-Butyl Ether (MTBE) 4. Ethyl *ter*-butyl ether (ETBE) 5. T*er*-amyl methyl ether (TAME) 6. *Ter*-amyl ethyl Ether (TAEE)

The European Fuel Oxygenates Association (aka EFOA) credits Methyl Tertiary-Butyl Ether (MTBE) and Ethyl ter-butyl ether (ETBE) as the most commonly used ethers in fuel to replace lead. Ethers were brought into fuels in Europe in the 1970s to replace the highly toxic compound. Although Europeans still use Bio-ether additives, the US no longer has an oxygenate requirement therefore bio-ethers are no longer used as the main fuel additive.

Biogas

Pipes carrying biogas

Biogas is methane produced by the process of anaerobic digestion of organic material by anaerobes. It can be produced either from biodegradable waste materials or by the use of energy crops fed into anaerobic digesters to supplement gas yields. The solid byproduct, digestate, can be used as a biofuel or a fertilizer.

- Biogas can be recovered from mechanical biological treatment waste processing systems.

- Note: Landfill gas, a less clean form of biogas, is produced in landfills through naturally occurring anaerobic digestion. If it escapes into the atmosphere, it is a potential greenhouse gas.

- Farmers can produce biogas from manure from their cattle by using anaerobic digesters.

Syngas

Syngas, a mixture of carbon monoxide, hydrogen and other hydrocarbons, is produced by partial combustion of biomass, that is, combustion with an amount of oxygen that is not sufficient to convert the biomass completely to carbon dioxide and water. Before partial combustion, the biomass is dried, and sometimes pyrolysed. The resulting gas mixture, syngas, is more efficient than direct combustion of the original biofuel; more of the energy contained in the fuel is extracted.

- Syngas may be burned directly in internal combustion engines, turbines or high-temperature fuel cells. The wood gas generator, a wood-fueled gasification reactor, can be connected to an internal combustion engine.

- Syngas can be used to produce methanol, DME and hydrogen, or converted via the Fischer-Tropsch process to produce a diesel substitute, or a mixture of alcohols that can be blended into gasoline. Gasification normally relies on temperatures greater than 700 °C.

- Lower-temperature gasification is desirable when co-producing biochar, but results in syngas polluted with tar.

Solid Biofuels

Examples include wood, sawdust, grass trimmings, domestic refuse, charcoal, agricultural waste, nonfood energy crops, and dried manure.

When raw biomass is already in a suitable form (such as firewood), it can burn directly in a stove or furnace to provide heat or raise steam. When raw biomass is in an inconvenient form (such as sawdust, wood chips, grass, urban waste wood, agricultural residues), the typical process is to densify the biomass. This process includes grinding the raw biomass to an appropriate particulate size (known as hogfuel), which, depending on the densification type, can be from 1 to 3 cm (0.4 to 1.2 in), which is then concentrated into a fuel product. The current processes produce wood pellets, cubes, or pucks. The pellet process is most common in Europe, and is typically a pure wood product. The other types of densification are larger in size compared to a pellet, and are compatible with a broad range of input feedstocks. The resulting densified fuel is easier to transport and feed into thermal generation systems, such as boilers.

Industry has used sawdust, bark and chips for fuel for decades, primary in the pulp and paper industry, and also bagasse (spent sugar cane) fueled boilers in the sugar cane industry. Boilers in the range of 500,000 lb/hr of steam, and larger, are in routine operation, using grate, spreader stoker, suspension burning and fluid bed combustion. Utilities generate power, typically in the range of 5 to 50 MW, using locally available fuel. Other industries have also installed wood waste fueled boilers and dryers in areas with low cost fuel.

One of the advantages of biomass fuel is that it is often a byproduct, residue or waste-product of other processes, such as farming, animal husbandry and forestry. In theory, this means fuel and food production do not compete for resources, although this is not always the case.

A problem with the combustion of raw biomass is that it emits considerable amounts of pollutants, such as particulates and polycyclic aromatic hydrocarbons. Even modern pellet boilers generate much more pollutants than oil or natural gas boilers. Pellets made from agricultural residues are usually worse than wood pellets, producing much larger emissions of dioxins and chlorophenols.

In spite of the above noted study, numerous studies have shown biomass fuels have significantly less impact on the environment than fossil based fuels. Of note is the US Department of Energy Laboratory, operated by Midwest Research Institute Biomass Power and Conventional Fossil Systems with and without CO2 Sequestration – Comparing the Energy Balance, Greenhouse Gas Emissions and Economics Study. Power generation emits significant amounts of greenhouse gases (GHGs), mainly carbon dioxide (CO_2). Sequestering CO_2 from the power plant flue gas can significantly reduce the GHGs from the power plant itself, but this is not the total picture. CO_2 capture and sequestration consumes additional energy, thus lowering the plant's fuel-to-electricity efficiency. To compensate for this, more fossil fuel must be procured and consumed to make up for lost capacity.

Taking this into consideration, the global warming potential (GWP), which is a combination of CO_2, methane (CH_4), and nitrous oxide (N_2O) emissions, and energy balance of the system need to be examined using a life cycle assessment. This takes into account the upstream processes which remain constant after CO_2 sequestration, as well as the steps required for additional power generation. Firing biomass instead of coal led to a 148% reduction in GWP.

A derivative of solid biofuel is biochar, which is produced by biomass pyrolysis. Biochar made from agricultural waste can substitute for wood charcoal. As wood stock becomes scarce, this alternative is gaining ground. In eastern Democratic Republic of Congo, for example, biomass briquettes are being marketed as an alternative to charcoal to protect Virunga National Park from deforestation associated with charcoal production.

Second-generation (Advanced) Biofuels

Second generation biofuels, also known as advanced biofuels, are fuels that can be manufactured from various types of biomass. Biomass is a wide-ranging term meaning any source of organic carbon that is renewed rapidly as part of the carbon cycle. Biomass is derived from plant materials but can also include animal materials.

First generation biofuels are made from the sugars and vegetable oils found in arable crops, which can be easily extracted using conventional technology. In comparison, second generation biofuels are made from lignocellulosic biomass or woody crops, agricultural residues or waste, which makes it harder to extract the required fuel. A series of physical and chemical treatments might be required to convert lignocellulosic biomass to liquid fuels suitable for transportation.

Sustainable Biofuels

Biofuels in the form of liquid fuels derived from plant materials, are entering the market, driven mainly by the perception that they reduce climate gas emissions, and also by factors such as oil price spikes and the need for increased energy security. However, many of the biofuels that are currently being supplied have been criticised for their adverse impacts on the natural environment, food security, and land use. In 2008, the Nobel-prize winning chemist Paul J. Crutzen published findings that the release of nitrous oxide (N_2O) emissions in the production of biofuels means that overall they contribute more to global warming than the fossil fuels they replace.

The challenge is to support biofuel development, including the development of new cellulosic technologies, with responsible policies and economic instruments to help ensure that biofuel commercialization is sustainable. Responsible commercialization of biofuels represents an opportunity to enhance sustainable economic prospects in Africa, Latin America and Asia.

According to the Rocky Mountain Institute, sound biofuel production practices would not hamper food and fibre production, nor cause water or environmental problems, and would enhance soil fertility. The selection of land on which to grow the feedstocks is a critical component of the ability of biofuels to deliver sustainable solutions. A key consideration is the minimisation of biofuel competition for prime cropland.

Biofuels by Region

Bio Diesel Powered Fast Attack Craft Of Indian Navy patrolling during IFR 2016.The green bands on the vessels are indicative of the fact that the vessels are powered by bio-diesel

There are international organizations such as IEA Bioenergy, established in 1978 by the OECD International Energy Agency (IEA), with the aim of improving cooperation and information exchange between countries that have national programs in bioenergy research, development and deployment. The UN International Biofuels Forum is formed by Brazil, China, India, Pakistan, South Africa, the United States and the European Commission. The world leaders in biofuel development and use are Brazil, the United States, France, Sweden and Germany. Russia also has 22%

of world's forest, and is a big biomass (solid biofuels) supplier. In 2010, Russian pulp and paper maker, Vyborgskaya Cellulose, said they would be producing pellets that can be used in heat and electricity generation from its plant in Vyborg by the end of the year. The plant will eventually produce about 900,000 tons of pellets per year, making it the largest in the world once operational.

Biofuels currently make up 3.1% of the total road transport fuel in the UK or 1,440 million litres. By 2020, 10% of the energy used in UK road and rail transport must come from renewable sources – this is the equivalent of replacing 4.3 million tonnes of fossil oil each year. Conventional biofuels are likely to produce between 3.7 and 6.6% of the energy needed in road and rail transport, while advanced biofuels could meet up to 4.3% of the UK's renewable transport fuel target by 2020.

Air Pollution

Biofuels are different from fossil fuels in regard to greenhouse gases but are similar to fossil fuels in that biofuels contribute to air pollution. Burning produces airborne carbon particulates, carbon monoxide and nitrous oxides. The WHO estimates 3.7 million premature deaths worldwide in 2012 due to air pollution. Brazil burns significant amounts of ethanol biofuel. Gas chromatograph studies were performed of ambient air in São Paulo, Brazil, and compared to Osaka, Japan, which does not burn ethanol fuel. Atmospheric Formaldehyde was 160% higher in Brazil, and Acetaldehyde was 260% higher.

Debates Regarding the Production and Use of Biofuel

There are various social, economic, environmental and technical issues with biofuel production and use, which have been discussed in the popular media and scientific journals. These include: the effect of moderating oil prices, the "food vs fuel" debate, food prices, poverty reduction potential, energy ratio, energy requirements, carbon emissions levels, sustainable biofuel production, deforestation and soil erosion, loss of biodiversity, impact on water resources, the possible modifications necessary to run the engine on biofuel, as well as energy balance and efficiency. The International Resource Panel, which provides independent scientific assessments and expert advice on a variety of resource-related themes, assessed the issues relating to biofuel use in its first report *Towards sustainable production and use of resources: Assessing Biofuels*. "Assessing Biofuels" outlined the wider and interrelated factors that need to be considered when deciding on the relative merits of pursuing one biofuel over another. It concluded that not all biofuels perform equally in terms of their impact on climate, energy security and ecosystems, and suggested that environmental and social impacts need to be assessed throughout the entire life-cycle.

Another issue with biofuel use and production is the US has changed mandates many times because the production has been taking longer than expected. The Renewable Fuel Standard (RFS) set by congress for 2010 was pushed back to at best 2012 to produce 100 million gallons of pure ethanol (not blended with a fossil fuel).

Current Research

Research is ongoing into finding more suitable biofuel crops and improving the oil yields of these crops. Using the current yields, vast amounts of land and fresh water would be needed to produce enough oil to completely replace fossil fuel usage. It would require twice the land area of the US to be devoted to soybean production, or two-thirds to be devoted to rapeseed production, to meet current US heating and transportation needs.

Specially bred mustard varieties can produce reasonably high oil yields and are very useful in crop rotation with cereals, and have the added benefit that the meal left over after the oil has been pressed out can act as an effective and biodegradable pesticide.

The NFESC, with Santa Barbara-based Biodiesel Industries, is working to develop biofuels technologies for the US navy and military, one of the largest diesel fuel users in the world. A group of Spanish developers working for a company called Ecofasa announced a new biofuel made from trash. The fuel is created from general urban waste which is treated by bacteria to produce fatty acids, which can be used to make biofuels.

Ethanol Biofuels

As the primary source of biofuels in North America, many organizations are conducting research in the area of ethanol production. The National Corn-to-Ethanol Research Center (NCERC) is a research division of Southern Illinois University Edwardsville dedicated solely to ethanol-based biofuel research projects. On the federal level, the USDA conducts a large amount of research regarding ethanol production in the United States. Much of this research is targeted toward the effect of ethanol production on domestic food markets. A division of the U.S. Department of Energy, the National Renewable Energy Laboratory (NREL), has also conducted various ethanol research projects, mainly in the area of cellulosic ethanol.

Cellulosic ethanol commercialization is the process of building an industry out of methods of turning cellulose-containing organic matter into fuel. Companies, such as Iogen, POET, and Abengoa, are building refineries that can process biomass and turn it into bioethanol. Companies, such as Diversa, Novozymes, and Dyadic, are producing enzymes that could enable a cellulosic ethanol future. The shift from food crop feedstocks to waste residues and native grasses offers significant opportunities for a range of players, from farmers to biotechnology firms, and from project developers to investors.

As of 2013, the first commercial-scale plants to produce cellulosic biofuels have begun operating. Multiple pathways for the conversion of different biofuel feedstocks are being used. In the next few years, the cost data of these technologies operating at commercial scale, and their relative performance, will become available. Lessons learnt will lower the costs of the industrial processes involved.

In parts of Asia and Africa where drylands prevail, sweet sorghum is being investigated as a potential source of food, feed and fuel combined. The crop is particularly suitable for growing in arid conditions, as it only extracts one seventh of the water used by sugarcane. In India, and other places, sweet sorghum stalks are used to produce biofuel by squeezing the juice and then fermenting into ethanol.

A study by researchers at the International Crops Research Institute for the Semi-Arid Tropics (ICRISAT) found that growing sweet sorghum instead of grain sorghum could increase farmers incomes by US$40 per hectare per crop because it can provide fuel in addition to food and animal feed. With grain sorghum currently grown on over 11 million hectares (ha) in Asia and on 23.4 million ha in Africa, a switch to sweet sorghum could have a considerable economic impact.

Algae Biofuels

From 1978 to 1996, the US NREL experimented with using algae as a biofuels source in the "Aquatic Species Program". A self-published article by Michael Briggs, at the UNH Biofuels Group, offers estimates for the realistic replacement of all vehicular fuel with biofuels by using algae that have a natural oil content greater than 50%, which Briggs suggests can be grown on algae ponds at wastewater treatment plants. This oil-rich algae can then be extracted from the system and processed into biofuels, with the dried remainder further reprocessed to create ethanol. The production of algae to harvest oil for biofuels has not yet been undertaken on a commercial scale, but feasibility studies have been conducted to arrive at the above yield estimate. In addition to its projected high yield, algaculture — unlike crop-based biofuels — does not entail a decrease in food production, since it requires neither farmland nor fresh water. Many companies are pursuing algae bioreactors for various purposes, including scaling up biofuels production to commercial levels. Prof. Rodrigo E. Teixeira from the University of Alabama in Huntsville demonstrated the extraction of biofuels lipids from wet algae using a simple and economical reaction in ionic liquids.

Jatropha

Several groups in various sectors are conducting research on *Jatropha curcas*, a poisonous shrub-like tree that produces seeds considered by many to be a viable source of biofuels feedstock oil. Much of this research focuses on improving the overall per acre oil yield of Jatropha through advancements in genetics, soil science, and horticultural practices.

SG Biofuels, a San Diego-based jatropha developer, has used molecular breeding and biotechnology to produce elite hybrid seeds that show significant yield improvements over first-generation varieties. SG Biofuels also claims additional benefits have arisen from such strains, including improved flowering synchronicity, higher resistance to pests and diseases, and increased cold-weather tolerance.

Plant Research International, a department of the Wageningen University and Research Centre in the Netherlands, maintains an ongoing Jatropha Evaluation Project that examines the feasibility of large-scale jatropha cultivation through field and laboratory experiments. The Center for Sustainable Energy Farming (CfSEF) is a Los Angeles-based nonprofit research organization dedicated to jatropha research in the areas of plant science, agronomy, and horticulture. Successful exploration of these disciplines is projected to increase jatropha farm production yields by 200-300% in the next 10 years.

Fungi

A group at the Russian Academy of Sciences in Moscow, in a 2008 paper, stated they had isolated large amounts of lipids from single-celled fungi and turned it into biofuels in an economically efficient manner. More research on this fungal species, *Cunninghamella japonica*, and others, is likely to appear in the near future. The recent discovery of a variant of the fungus *Gliocladium roseum* (later renamed Ascocoryne sarcoides) points toward the production of so-called myco-diesel from cellulose. This organism was recently discovered in the rainforests of northern Patagonia, and has the unique capability of converting cellulose into medium-length hydrocarbons typically found in diesel fuel. Many other fungi that can degrade cellulose and other polymers have been observed to produce molecules that are currently being engineered using organisms from other kingdoms, suggesting that fungi may play a large role in the bio-production of fuels in the future (reviewed in).

Animal Gut Bacteria

Microbial gastrointestinal flora in a variety of animals have shown potential for the production of biofuels. Recent research has shown that TU-103, a strain of *Clostridium* bacteria found in Zebra feces, can convert nearly any form of cellulose into butanol fuel. Microbes in panda waste are being investigated for their use in creating biofuels from bamboo and other plant materials. There has also been substantial research into the technology of using the gut microbiomes of wood-feeding insects for the conversion of lignocellulotic material into biofuel.

Greenhouse Gas Emissions

Some scientists have expressed concerns about land-use change in response to greater demand for crops to use for biofuel and the subsequent carbon emissions. The payback period, that is, the time it will take biofuels to pay back the carbon debt they acquire due to land-use change, has been estimated to be between 100 and 1000 years, depending on the specific instance and location of land-use change. However, no-till practices combined with cover-crop practices can reduce the payback period to three years for grassland conversion and 14 years for forest conversion.

A study conducted in the Tocantis State, in northern Brazil, found that many families were cutting down forests in order to produce two conglomerates of oilseed plants, the J. curcas (JC group) and the R. communis (RC group). This region is composed of 15% Amazonian rainforest with high biodiversity, and 80% Cerrado forest with lower biodiversity. During the study, the farmers that planted the JC group released over 2193 Mg CO_2, while losing 53-105 Mg CO_2 sequestration from deforestation; and the RC group farmers released 562 Mg CO_2, while losing 48-90 Mg CO_2 to be sequestered from forest depletion. The production of these types of biofuels not only led into an increased emission of carbon dioxide, but also to lower efficiency of forests to absorb the gases that these farms were emitting. This has to do with the amount of fossil fuel the production of fuel crops involves. In addition, the intensive use of monocropping agriculture requires large amounts of water irrigation, as well as of fertilizers, herbicides and pesticides. This does not only lead to poisonous chemicals to disperse on water runoff, but also to the emission of nitrous oxide (NO_2) as a fertilizer byproduct, which is three hundred times more efficient in producing a greenhouse effect than carbon dioxide (CO_2).

Converting rainforests, peatlands, savannas, or grasslands to produce food crop–based biofuels in Brazil, Southeast Asia, and the United States creates a "biofuel carbon debt" by releasing 17 to 420 times more CO_2 than the annual greenhouse gas (GHG) reductions that these biofuels would provide by displacing fossil fuels. Biofuels made from waste biomass or from biomass grown on abandoned agricultural lands incur little to no carbon debt.

Water Use

In addition to water required to grow crops, biofuel facilities require significant process water.

References

- Spakowicz, Daniel J.; Strobel, Scott A. (2015). "Biosynthesis of hydrocarbons and volatile organic compounds by fungi: bioengineering potential". Applied microbiology and biotechnology. 99 (12): 4943–4951. Retrieved 2016-02-22.

- The National Academies Press (2008). "Water Issues of Biofuel Production Plants". The National Academies Press. Retrieved 18 June 2015.

- REN21 (2011). "Renewables 2011: Global Status Report" (PDF). pp. 13–14. Archived from the original (PDF) on 2011-09-05. Retrieved 2015-01-03.

- "The potential and challenges of drop-in fuels (members only) | IEA Bioenergy Task 39 – Commercializing Liquid Biofuels". task39.sites.olt.ubc.ca. Retrieved 2015-09-10.

- Cotton, Charles A. R.; Jeffrey S. Douglass; Sven De Causmaeker; Katharina Brinkert; Tanai Cardona; Andrea Fantuzzi; A. William Rutherford; James W. Murray (2015). "Photosynthetic constraints on fuel from microbes". Frontiers in Bioengineering and Biotechnology. 3. doi:10.3389/fbioe.2015.00036. Retrieved 18 March 2015.

- Fletcher Jr., Robert J.; Bruce A Robertson; Jason Evans; Patrick J Doran; Janaki RR Alavalapati; Douglas W Schemske (2011). "Biodiversity conservation in the era of biofuels: risks and opportunities". Frontiers in Ecology and the Environment. 9 (3): 161–168. doi:10.1890/090091. Retrieved 10 December 2013.

- Ethanol Research (2012-04-02). "National Corn-to-Ethanol Research Center (NCERC)". Ethanol Research. Archived from the original on 20 March 2012. Retrieved 2012-04-02.

- American Coalition for Ethanol (2008-06-02). "Responses to Questions from Senator Bingaman" (PDF). American Coalition for Ethanol. Archived from the original (PDF) on 4 October 2011. Retrieved 2012-04-02.

- National Renewable Energy Laboratory (2007-03-02). "Research Advantages: Cellulosic Ethanol" (PDF). National Renewable Energy Laboratory. Retrieved 2012-04-02.

- Sheehan, John; et al. (July 1998). "A Look Back at the U. S. Department of Energy's Aquatic Species Program: Biofuels from Algae" (PDF). National Renewable Energy Laboratory. Retrieved 16 June 2012.

- Biofuels Digest (2011-05-16). "Jatropha blooms again: SG Biofuels secures 250K acres for hybrids". Biofuels Digest. Retrieved 2012-03-08.

- Plant Research International (2012-03-08). "JATROPT (Jatropha curcas): Applied and technical research into plant properties". Plant Research International. Retrieved 2012-03-08.

- Kathryn Hobgood Ray (August 25, 2011). "Cars Could Run on Recycled Newspaper, Tulane Scientists Say". Tulane University news webpage. Tulane University. Retrieved March 14, 2012.

- National Non-Food Crops Centre. "Advanced Biofuels: The Potential for a UK Industry, NNFCC 11-011", Retrieved on 2011-11-17

- "Mustard Hybrids for Low-Cost Biofuels and Organic Pesticides" (PDF). Archived from the original (PDF) on 26 July 2011. Retrieved 2010-03-15.

- Searchinger, Timothy; Ralph Heimlich; R.A. Houghton; Fengxia Dong; Amani Elobeid; Jacinto Fabiosa; Simla Tokgoz; Dermot Hayes; Tun-Hsiang Yu (2011). "Use of U.S. Croplands for Biofuels Increases Greenhouse Gases Through Emissions from Land-Use Change". Science. pp. 1238–1240. doi:10.1126/science.1151861. Retrieved 8 November 2011.

- fargione, Joseph; Jason Hill; David Tilman; Stephen Polasky; Peter Hawthorne (2008). "Land Clearing and the Biofuel Carbon Debt". Science. pp. 1235–1238. doi:10.1126/science.1152747. Retrieved 12 November 2011.

Biofuels and its Types

The types of biofuels discussed in this chapter are ethanol fuel, algae fuel, biodiesel and biogas. The most common biofuel worldwide is the ethanol fuel. While ethanol is particularly used in Brazil, one of the other types of biofuels, biodiesel is widely used in Europe. The topics discussed in the chapter are of great importance to broaden the existing knowledge on biofuels.

Ethanol Fuel

Ethanol fuel is ethyl alcohol, the same type of alcohol found in alcoholic beverages, used as fuel. It is most often used as a motor fuel, mainly as a biofuel additive for gasoline. The first production car running entirely on ethanol was the Fiat 147, introduced in 1978 in Brazil by Fiat. Nowadays, cars are able to run using 100% ethanol fuel or a mix of ethanol and gasoline (aka flex-fuel). It is commonly made from biomass such as corn or sugarcane. World ethanol production for transport fuel tripled between 2000 and 2007 from 17 billion to more than 52 billion liters. From 2007 to 2008, the share of ethanol in global gasoline type fuel use increased from 3.7% to 5.4%. In 2011 worldwide ethanol fuel production reached 22.36 billion U.S. liquid gallons (bg) (84.6 billion liters), with the United States as the top producer with 13.9 bg (52.6 billion liters), accounting for 62.2% of global production, followed by Brazil with 5.6 bg (21.1 billion liters). Ethanol fuel has a "gasoline gallon equivalency" (GGE) value of 1.5 US gallons (5.7 L), which means 1.5 gallons of ethanol produces the energy of one gallon of gasoline.

The Saab 9-3 SportCombi BioPower was the second E85 flexifuel model introduced by Saab in the Swedish market.

Ethanol fuel is widely used in Brazil and in the United States, and together both countries were responsible for 87.1% of the world's ethanol fuel production in 2011. Most cars on the road today in the U.S. can run on blends of up to 10% ethanol, and ethanol represented 10% of the U.S. gasoline fuel supply derived from domestic sources in 2011. Since 1976 the Brazilian government has made it mandatory to blend ethanol with gasoline, and since 2007 the legal blend is around 25% ethanol and 75% gasoline (E25). By December 2011 Brazil had a fleet of 14.8 million flex-fuel automobiles and light trucks and 1.5 million flex-fuel motorcycles that regularly use neat ethanol fuel (known as E100).

Bioethanol is a form of quasi-renewable energy that can be produced from agricultural feedstocks. It can be made from very common crops such as hemp, sugarcane, potato, cassava and corn. There has been considerable debate about how useful bioethanol is in replacing gasoline. Concerns about its production and use relate to increased food prices due to the large amount of arable land required for crops, as well as the energy and pollution balance of the whole cycle of ethanol production, especially from corn. Recent developments with cellulosic ethanol production and commercialization may allay some of these concerns.

Cellulosic ethanol offers promise because cellulose fibers, a major and universal component in plant cells walls, can be used to produce ethanol. According to the International Energy Agency, cellulosic ethanol could allow ethanol fuels to play a much bigger role in the future.

Chemistry

Structure of ethanol molecule. All bonds are single bonds

During ethanol fermentation, glucose and other sugars in the corn (or sugarcane or other crops) are converted into ethanol and carbon dioxide.

$$C_6H_{12}O_6 \rightarrow 2\ C_2H_5OH + 2\ CO_2 + heat$$

Ethanol fermentation is not 100% selective with other side products such as acetic acid, glycols and many other products produced. They are mostly removed during ethanol purification. Fermentation takes place in an aqueous solution. The resulting solution has an ethanol content of around 15%. Ethanol is subsequently isolated and purified by a combination of adsorption and distillation.

During combustion, ethanol reacts with oxygen to produce carbon dioxide, water, and heat:

$$C_2H_5OH + 3\ O_2 \rightarrow 2\ CO_2 + 3\ H_2O + heat$$

Starch and cellulose molecules are strings of glucose molecules. It is also possible to generate ethanol out of cellulosic materials. That, however, requires a pretreatment that splits the cellulose into glycose molecules and other sugars that subsequently can be fermented. The resulting product is called cellulosic ethanol, indicating its source.

Ethanol may also be produced industrially from ethylene by hydration of the double bond in the presence of catalysts and high temperature.

$$C_2H_4 + H_2O \rightarrow C_2H_5OH$$

By far the largest fraction of the global ethanol production, however, is produced by fermentation.

Sources

Sugar cane harvest

Cornfield in South Africa

Switchgrass

Ethanol is a quasi-renewable energy source because while the energy is partially generated by using a resource, sunlight, which cannot be depleted, the harvesting process requires vast amounts of energy that typically comes from non-renewable sources. Creation of ethanol starts with photosynthesis causing a feedstock, such as sugar cane or a grain such as maize (corn), to grow. These feedstocks are processed into ethanol.

About 5% of the ethanol produced in the world in 2003 was actually a petroleum product. It is made by the catalytic hydration of ethylene with sulfuric acid as the catalyst. It can also be obtained via ethylene or acetylene, from calcium carbide, coal, oil gas, and other sources. Two million tons of petroleum-derived ethanol are produced annually. The principal suppliers are plants in the United States, Europe, and South Africa. Petroleum derived ethanol (synthetic ethanol) is chemically identical to bio-ethanol and can be differentiated only by radiocarbon dating.

Bio-ethanol is usually obtained from the conversion of carbon-based feedstock. Agricultural feedstocks are considered renewable because they get energy from the sun using photosynthesis, provided that all minerals required for growth (such as nitrogen and phosphorus) are returned to the land. Ethanol can be produced from a variety of feedstocks such as sugar cane, bagasse, miscanthus, sugar beet, sorghum, grain, switchgrass, barley, hemp, kenaf, potatoes, sweet potatoes, cassava, sunflower, fruit, molasses, corn, stover, grain, wheat, straw, cotton, other biomass, as well as many types of cellulose waste and harvesting, whichever has the best well-to-wheel assessment.

An alternative process to produce bio-ethanol from algae is being developed by the company Algenol. Rather than grow algae and then harvest and ferment it, the algae grow in sunlight and produce ethanol directly, which is removed without killing the algae. It is claimed the process can produce 6,000 US gallons per acre (56,000 litres per ha) per year compared with 400 US gallons per acre (3,750 l/ha) for corn production.

Currently, the first generation processes for the production of ethanol from corn use only a small part of the corn plant: the corn kernels are taken from the corn plant and only the starch, which represents about 50% of the dry kernel mass, is transformed into ethanol. Two types of second generation processes are under development. The first type uses enzymes and yeast fermentation to convert the plant cellulose into ethanol while the second type uses pyrolysis to convert the whole plant to either a liquid bio-oil or a syngas. Second generation processes can also be used with plants such as grasses, wood or agricultural waste material such as straw.

Production

The basic steps for large-scale production of ethanol are: microbial (yeast) fermentation of sugars, distillation, dehydration, and denaturing (optional). Prior to fermentation, some crops require saccharification or hydrolysis of carbohydrates such as cellulose and starch into sugars. Saccharification of cellulose is called cellulolysis. Enzymes are used to convert starch into sugar.

Fermentation

Ethanol is produced by microbial fermentation of the sugar. Microbial fermentation currently only works directly with sugars. Two major components of plants, starch and cellulose, are both made of sugars—and can, in principle, be converted to sugars for fermentation. Currently, only the sugar (e.g., sugar cane) and starch (e.g., corn) portions can be economically converted. There is much activity in the area of cellulosic ethanol, where the cellulose part of a plant is broken down to sugars and subsequently converted to ethanol.

Distillation

Ethanol plant in West Burlington, Iowa

Ethanol plant in Sertãozinho, Brazil.

For the ethanol to be usable as a fuel, the majority of the water must be removed. Most of the water is removed by distillation, but the purity is limited to 95–96% due to the formation of a low-boiling water-ethanol azeotrope with maximum (95.6% m/m (96.5% v/v) ethanol and 4.4% m/m (3.5% v/v) water). This mixture is called hydrous ethanol and can be used as a fuel alone, but unlike anhydrous ethanol, hydrous ethanol is not miscible in all ratios with gasoline, so the water fraction is typically removed in further treatment to burn in combination with gasoline in gasoline engines.

Dehydration

There are basically three dehydration processes to remove the water from an azeotropic ethanol/water mixture. The first process, used in many early fuel ethanol plants, is called azeotropic distillation and consists of adding benzene or cyclohexane to the mixture. When these components are added to the mixture, it forms a heterogeneous azeotropic mixture in vapor–liquid-liquid equilibrium, which when distilled produces anhydrous ethanol in the column bottom, and a vapor mixture of water, ethanol, and cyclohexane/benzene.

When condensed, this becomes a two-phase liquid mixture. The heavier phase, poor in the entrainer (benzene or cyclohexane), is stripped of the entrainer and recycled to the feed—while the lighter phase, with condensate from the stripping, is recycled to the second column. Another early method, called extractive distillation, consists of adding a ternary component that increases ethanol's relative volatility. When the ternary mixture is distilled, it produces anhydrous ethanol on the top stream of the column.

With increasing attention being paid to saving energy, many methods have been proposed that avoid distillation altogether for dehydration. Of these methods, a third method has emerged and has been adopted by the majority of modern ethanol plants. This new process uses molecular sieves to remove water from fuel ethanol. In this process, ethanol vapor under pressure passes through a bed of molecular sieve beads. The bead's pores are sized to allow absorption of water while excluding ethanol. After a period of time, the bed is regenerated under vacuum or in the flow of inert atmosphere (e.g. N_2) to remove the absorbed water. Two beds are often used so that one is available to absorb water while the other is being regenerated. This dehydration technology can account for energy saving of 3,000 btus/gallon (840 kJ/L) compared to earlier azeotropic distillation.

Post-production Water Issues

Ethanol is hygroscopic, meaning it absorbs water vapor directly from the atmosphere. Because absorbed water dilutes the fuel value of the ethanol and may cause phase separation of ethanol-gasoline blends (which causes engine stall), containers of ethanol fuels must be kept tightly sealed. This high miscibility with water means that ethanol cannot be efficiently shipped through modern pipelines, like liquid hydrocarbons, over long distances. Mechanics also have seen increased cases of damage to small engines, in particular, the carburetor, attributable to the increased water retention by ethanol in fuel.

The fraction of water that an ethanol-gasoline fuel can contain without phase separation increases with the percentage of ethanol. This shows, for example, that E30 can have up to about 2% water. If there is more than about 71% ethanol, the remainder can be any proportion of water or gasoline and phase separation does not occur. The fuel mileage declines with increased water content. The increased solubility of water with higher ethanol content permits E30 and hydrated ethanol to be put in the same tank since any combination of them always results in a single phase. Somewhat less water is tolerated at lower temperatures. For E10 it is about 0.5% v/v at 70 F and decreases to about 0.23% v/v at −30 F.

Consumer Production Systems

While biodiesel production systems have been marketed to home and business users for many years, commercialized ethanol production systems designed for end-consumer use have lagged in the marketplace. In 2008, two different companies announced home-scale ethanol production systems. The AFS125 Advanced Fuel System from Allard Research and Development is capable of producing both ethanol and biodiesel in one machine, while the E-100 MicroFueler from E-Fuel Corporation is dedicated to ethanol only.

Engines

Ethanol most commonly powers Otto cycle internal combustion engines, most often on cars. However, it may be used to power vehicles using a Diesel cycle such as buses and farm tractors. Ethanol has been tested for use as aviation fuel but it is not commercialised .

Fuel Economy

Ethanol contains approx. 34% less energy per unit volume than gasoline, and therefore in theory, burning pure ethanol in a vehicle reduces miles per US gallon 34%, given the same fuel economy, compared to burning pure gasoline. However, since ethanol has a higher octane rating, the engine can be made more efficient by raising its compression ratio. Using a variable turbocharger, the compression ratio can be optimized for the fuel, making fuel economy almost constant for any blend.

For E10 (10% ethanol and 90% gasoline), the effect is small (~3%) when compared to conventional gasoline, and even smaller (1–2%) when compared to oxygenated and reformulated blends. For E85 (85% ethanol), the effect becomes significant. E85 produces lower mileage than gasoline, and requires more frequent refueling. Actual performance may vary depending on the vehicle. Based on EPA tests for all 2006 E85 models, the average fuel economy for E85 vehicles resulted 25.56% lower than unleaded gasoline. The EPA-rated mileage of current United States flex-fuel vehicles should be considered when making price comparisons, but E85 is a high performance fuel, with an octane rating of about 94–96, and should be compared to premium.

Cold Start During The Winter

The Brazilian 2008 Honda Civic flex-fuel has outside direct access to the secondary reservoir gasoline tank in the front right side, the corresponding fuel filler door is shown by the arrow.

High ethanol blends present a problem to achieve enough vapor pressure for the fuel to evaporate and spark the ignition during cold weather (since ethanol tends to increase fuel enthalpy of vaporization). When vapor pressure is below 45 kPa starting a cold engine becomes difficult. To avoid this problem at temperatures below 11 °C (52 °F), and to reduce ethanol higher emissions during cold weather, both the US and the European markets adopted E85 as the maximum blend to be used in their flexible fuel vehicles, and they are optimized to run at such a blend. At places with harsh cold weather, the ethanol blend in the US has a seasonal reduction to E70 for these very cold regions, though it is still sold as E85. At places where temperatures fall below –12 °C (10 °F) during the winter, it is recommended to install an engine heater system, both for gasoline and E85 vehicles. Sweden has a similar seasonal reduction, but the ethanol content in the blend is reduced to E75 during the winter months.

Brazilian flex fuel vehicles can operate with ethanol mixtures up to E100, which is hydrous ethanol (with up to 4% water), which causes vapor pressure to drop faster as compared to E85 vehicles. As a result, Brazilian flex vehicles are built with a small secondary gasoline reservoir located near the engine. During a cold start pure gasoline is injected to avoid starting problems at low temperatures. This provision is particularly necessary for users of Brazil's southern and central regions, where temperatures normally drop below 15 °C (59 °F) during the winter. An improved flex engine generation was launched in 2009 that eliminates the need for the secondary gas storage tank. In March 2009 Volkswagen do Brasil launched the Polo E-Flex, the first Brazilian flex fuel model without an auxiliary tank for cold start.

Fuel Mixtures

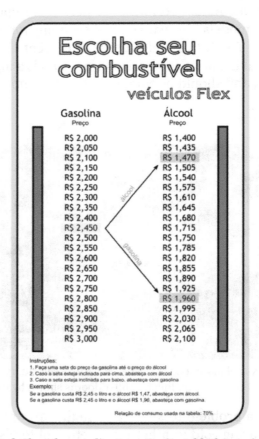

Hydrated ethanol × gasoline type C price table for use in Brazil

EPA's E15 label required to be displayed in all E15 fuel dispensers in the U.S.

In many countries cars are mandated to run on mixtures of ethanol. All Brazilian light-duty vehicles are built to operate for an ethanol blend of up to 25% (E25), and since 1993 a federal law requires mixtures between 22% and 25% ethanol, with 25% required as of mid July 2011. In the United States all light-duty vehicles are built to operate normally with an ethanol blend of 10% (E10). At the end of 2010 over 90 percent of all gasoline sold in the U.S. was blended with ethanol. In January 2011 the U.S. Environmental Protection Agency (EPA) issued a waiver to authorize up to 15% of ethanol blended with gasoline (E15) to be sold only for cars and light pickup trucks with a model year of 2001 or newer. Other countries have adopted their own requirements.

Beginning with the model year 1999, an increasing number of vehicles in the world are manufactured with engines that can run on any fuel from 0% ethanol up to 100% ethanol without modification. Many cars and light trucks (a class containing minivans, SUVs and pickup trucks) are designed to be flexible-fuel vehicles using ethanol blends up to 85% (E85) in North America and Europe, and up to 100% (E100) in Brazil. In older model years, their engine systems contained alcohol sensors in the fuel and/or oxygen sensors in the exhaust that provide input to the engine control computer to adjust the fuel injection to achieve stochiometric (no residual fuel or free oxygen in the exhaust) air-to-fuel ratio for any fuel mix. In newer models, the alcohol sensors have been removed, with the computer using only oxygen and airflow sensor feedback to estimate alcohol content. The engine control computer can also adjust (advance) the ignition timing to achieve a higher output without pre-ignition when it predicts that higher alcohol percentages are present in the fuel being burned. This method is backed up by advanced knock sensors – used in most high performance gasoline engines regardless of whether they are designed to use ethanol or not – that detect pre-ignition and detonation.

Corrosion

Whether ethanol causes unacceptable levels of corrosion in internal combustion engines has been a cause of some debate. In general, E10 blends pose no problems. At higher blends, ethanol's different chemistry from oil derivates makes it corrosive for certain engine parts. These are replaced to create the flex-fuel vehicles that accept up to E85.

Other Engine Configurations

ED95 engines

Since 1989 there have also been ethanol engines based on the diesel principle operating in Sweden. They are used primarily in city buses, but also in distribution trucks and waste collectors. The engines, made by Scania, have a modified compression ratio, and the fuel (known as ED95) used is a mix of 93.6% ethanol and 3.6% ignition improver, and 2.8% denaturants. The ignition improver makes it possible for the fuel to ignite in the diesel combustion cycle. It is then also possible to use the energy efficiency of the diesel principle with ethanol. These engines have been used in the United Kingdom by Reading Transport but the use of bioethanol fuel is now being phased out.

Dual-fuel direct-injection

A 2004 MIT study and an earlier paper published by the Society of Automotive Engineers identified a method to exploit the characteristics of fuel ethanol substantially more efficiently than mixing it with gasoline. The method presents the possibility of leveraging the use of alcohol to achieve definite improvement over the cost-effectiveness of hybrid electric. The improvement consists of using dual-fuel direct-injection of pure alcohol (or the azeotrope or E85) and gasoline, in any ratio up to 100% of either, in a turbocharged, high compression-ratio, small-displacement engine having performance similar to an engine having twice the displacement. Each fuel is carried separately, with a much smaller tank for alcohol. The high-compression (for higher efficiency) engine runs on ordinary gasoline under low-power cruise conditions. Alcohol is directly injected into the cylinders (and the gasoline injection simultaneously reduced) only when necessary to suppress 'knock' such as when significantly accelerating. Direct cylinder injection raises the already high octane rating of ethanol up to an effective 130. The calculated over-all reduction of gasoline use and CO_2 emission is 30%. The consumer cost payback time shows a 4:1 improvement over turbo-diesel and a 5:1 improvement over hybrid. The problems of water absorption into pre-mixed gasoline (causing phase separation), supply issues of multiple mix ratios and cold-weather starting are also avoided.

Increased thermal efficiency

In a 2008 study, complex engine controls and increased exhaust gas recirculation allowed a compression ratio of 19.5 with fuels ranging from neat ethanol to E50. Thermal efficiency up to approximately that for a diesel was achieved. This would result in the fuel economy of a neat ethanol vehicle to be about the same as one burning gasoline.

Fuel cells powered by an ethanol reformer

In June 2016, Nissan announced plans to develop fuel cell vehicles powered by ethanol rather than hydrogen, the fuel of choice by the other car manufacturers that have developed and commercialized fuel cell vehicles, such as the Hyundai Tucson FCEV, Toyota Mirai, and Honda FCX Clarity. The main advantage of this technical approach is that it would be cheaper and easier to deploy the fueling infrastructure than setting up the one required to deliver hydrogen at high pressures, as each hydrogen fueling station cost US$1 million to US$2 million to build.

Nissan plans to create a technology that uses liquid ethanol fuel as a source to generate hydrogen within the vehicle itself. The technology uses heat to reform ethanol into hydrogen to feed what is known as a solid oxide fuel cell (SOFC). The fuel cell generates electricity to supply power to the electric motor driving the wheels, through a battery that handles peak power demands and stores regenerated energy. The vehicle would include a tank for a blend of water and ethanol, which is fed into an onboard reformer that splits it into pure hydrogen and carbon dioxide. According to Nissan, the liquid fuel could be an ethanol-water blend at a 55:45 ratio. Nissan expects to commercialize its technology by 2020.

Experience by Country

The world's top ethanol fuel producers in 2011 were the United States with 13.9 billion U.S. liquid gallons (bg) (52.60 billion liters) and Brazil with 5.6 bg (21.1 billion liters), accounting together for 87.1% of world production of 22.36 billion US gallons (84.6 billion liters). Strong incentives, coupled with other industry development initiatives, are giving rise to fledgling ethanol industries in countries such as Germany, Spain, France, Sweden, China, Thailand, Canada, Colombia, India, Australia, and some Central American countries.

Energy Balance

Energy balance		
Country	**Type**	**Energy balance**
United States	Corn ethanol	1.3
Germany	Biodiesel	2.5
Brazil	Sugarcane ethanol	8
United States	Cellulosic ethanol[†]	2–36[††]

[†] experimental, not in commercial production

[††] depending on production method

All biomass goes through at least some of these steps: it needs to be grown, collected, dried, fermented, distilled, and burned. All of these steps require resources and an infrastructure. The total amount of energy input into the process compared to the energy released by burning the resulting ethanol fuel is known as the energy balance (or "energy returned on energy invested"). Figures compiled in a 2007 by *National Geographic Magazine* point to modest results for corn ethanol produced in the US: one unit of fossil-fuel energy is required to create 1.3 energy units from the resulting ethanol. The energy balance for sugarcane ethanol produced in Brazil is more favorable, with one unit of fossil-fuel energy required to create 8 from the ethanol. Energy balance estimates are not easily produced, thus numerous such reports have been generated that are contradictory. For instance, a separate survey reports that production of ethanol from sugarcane, which requires a tropical climate to grow productively, returns from 8 to 9 units of energy for each unit expended, as compared to corn, which only returns about 1.34 units of fuel energy for each unit of energy expended. A 2006 University of California Berkeley study, after analyzing six separate studies, concluded that producing ethanol from corn uses much less petroleum than producing gasoline.

Carbon dioxide, a greenhouse gas, is emitted during fermentation and combustion. This is canceled out by the greater uptake of carbon dioxide by the plants as they grow to produce the biomass. When compared to gasoline, depending on the production method, ethanol releases less greenhouse gases.

Air Pollution

Compared with conventional unleaded gasoline, ethanol is a particulate-free burning fuel source that combusts with oxygen to form carbon dioxide, carbon monoxide, water and aldehydes. The Clean Air Act requires the addition of oxygenates to reduce carbon monoxide emissions in the United States. The additive MTBE is currently being phased out due to ground water contamination, hence ethanol becomes an attractive alternative additive. Current production methods include air pollution from the manufacturer of macronutrient fertilizers such as ammonia.

A study by atmospheric scientists at Stanford University found that E85 fuel would increase the risk of air pollution deaths relative to gasoline by 9% in Los Angeles, US: a very large, urban, car-based metropolis that is a worst-case scenario. Ozone levels are significantly increased, thereby increasing photochemical smog and aggravating medical problems such as asthma.

Brazil burns significant amounts of ethanol biofuel. Gas chromatograph studies were performed of ambient air in São Paulo, Brazil, and compared to Osaka, Japan, which does not burn ethanol fuel. Atmospheric Formaldehyde was 160% higher in Brazil, and Acetaldehyde was 260% higher.

Manufacture

In 2002, monitoring the process of ethanol production from corn revealed that they released VOCs (volatile organic compounds) at a higher rate than had previously been disclosed. The U.S. Environmental Protection Agency (EPA) subsequently reached settlement with Archer Daniels Midland and Cargill, two of the largest producers of ethanol, to reduce emission of these VOCs. VOCs are produced when fermented corn mash is dried for sale as a supplement for livestock feed. Devices known as thermal oxidizers or catalytic oxidizers can be attached to the plants to burn off the hazardous gases.

Carbon Dioxide

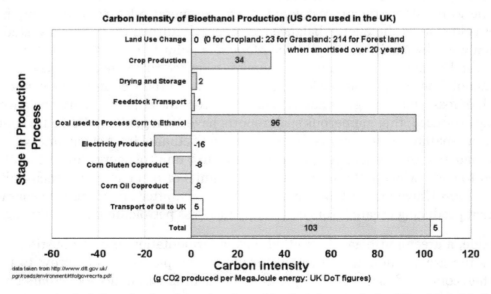

UK government calculation of carbon intensity of corn bioethanol grown in the US and burnt in the UK.

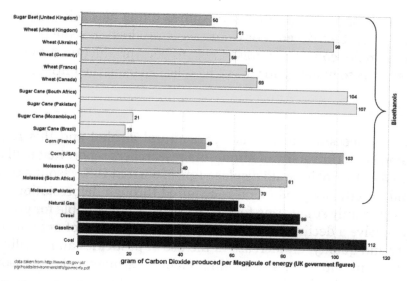

Graph of UK figures for the carbon intensity of bioethanol and fossil fuels. This graph assumes that all bioethanols are burnt in their country of origin and that previously existing cropland is used to grow the feedstock.

The calculation of exactly how much carbon dioxide is produced in the manufacture of bioethanol is a complex and inexact process, and is highly dependent on the method by which the ethanol is produced and the assumptions made in the calculation. A calculation should include:

- The cost of growing the feedstock
- The cost of transporting the feedstock to the factory
- The cost of processing the feedstock into bioethanol

Such a calculation may or may not consider the following effects:

- The cost of the change in land use of the area where the fuel feedstock is grown.
- The cost of transportation of the bioethanol from the factory to its point of use
- The efficiency of the bioethanol compared with standard gasoline
- The amount of Carbon Dioxide produced at the tail pipe.
- The benefits due to the production of useful bi-products, such as cattle feed or electricity.

The graph on the right shows figures calculated by the UK government for the purposes of the Renewable transport fuel obligation.

The January 2006 Science article from UC Berkeley's ERG, estimated reduction from corn ethanol in GHG to be 13% after reviewing a large number of studies. In a correction to that article released shortly after publication, they reduce the estimated value to 7.4%. A National Geographic Magazine overview article (2007) puts the figures at 22% less CO_2 emissions in production and use for corn ethanol compared to gasoline and a 56% reduction for cane ethanol. Carmaker Ford reports a 70% reduction in CO_2 emissions with bioethanol compared to petrol for one of their flexible-fuel vehicles.

An additional complication is that production requires tilling new soil which produces a one-off release of GHG that it can take decades or centuries of production reductions in GHG emissions to equalize. As an example, converting grass lands to corn production for ethanol takes about a century of annual savings to make up for the GHG released from the initial tilling.

Change in Land Use

Large-scale farming is necessary to produce agricultural alcohol and this requires substantial amounts of cultivated land. University of Minnesota researchers report that if all corn grown in the U.S. were used to make ethanol it would displace 12% of current U.S. gasoline consumption. There are claims that land for ethanol production is acquired through deforestation, while others have observed that areas currently supporting forests are usually not suitable for growing crops. In any case, farming may involve a decline in soil fertility due to reduction of organic matter, a decrease in water availability and quality, an increase in the use of pesticides and fertilizers, and potential dislocation of local communities. New technology enables farmers and processors to increasingly produce the same output using less inputs.

Cellulosic ethanol production is a new approach that may alleviate land use and related concerns. Cellulosic ethanol can be produced from any plant material, potentially doubling yields, in an effort to minimize conflict between food needs vs. fuel needs. Instead of utilizing only the starch by-products from grinding wheat and other crops, cellulosic ethanol production maximizes the use of all plant materials, including gluten. This approach would have a smaller carbon footprint because the amount of energy-intensive fertilisers and fungicides remain the same for higher output of usable material. The technology for producing cellulosic ethanol is currently in the commercialization stage.

Using Biomass for Electricity Instead of Ethanol

Converting biomass to electricity for charging electric vehicles may be a more "climate-friendly" transportation option than using biomass to produce ethanol fuel, according to an analysis published in Science in May 2009 Researchers continue to search for more cost-effective developments in both cellulosic ethanol and advanced vehicle batteries.

Health Costs of Ethanol Emissions

For each billion ethanol-equivalent gallons of fuel produced and combusted in the US, the combined climate-change and health costs are $469 million for gasoline, $472–952 million for corn ethanol depending on biorefinery heat source (natural gas, corn stover, or coal) and technology, but only $123–208 million for cellulosic ethanol depending on feedstock (prairie biomass, Miscanthus, corn stover, or switchgrass).

Efficiency of Common Crops

As ethanol yields improve or different feedstocks are introduced, ethanol production may become more economically feasible in the US. Currently, research on improving ethanol yields from each unit of corn is underway using biotechnology. Also, as long as oil prices remain high, the economical use of other feedstocks, such as cellulose, become viable. By-products such as straw or wood chips can be converted to ethanol. Fast growing species like switchgrass can be grown on land not suitable for other cash crops and yield high levels of ethanol per unit area.

Crop	Annual yield (Liters/hectare, US gal/acre)	Greenhouse-gas savings vs. petrol[a]	Comments
Sugar cane	6800–8000 L/ha, 727–870 g/acre	87%–96%	Long-season annual grass. Used as feedstock for most bio-ethanol produced in Brazil. Newer processing plants burn residues not used for ethanol to generate electricity. Grows only in tropical and subtropical climates.
Miscanthus	7300 L/ha, 780 g/acre	37%–73%	Low-input perennial grass. Ethanol production depends on development of cellulosic technology.
Switchgrass	3100–7600 L/ha, 330–810 g/acre	37%–73%	Low-input perennial grass. Ethanol production depends on development of cellulosic technology. Breeding efforts underway to increase yields. Higher biomass production possible with mixed species of perennial grasses.
Poplar	3700–6000 L/ha, 400–640 g/acre	51%–100%	Fast-growing tree. Ethanol production depends on development of cellulosic technology. Completion of genomic sequencing project will aid breeding efforts to increase yields.
Sweet sorghum	2500–7000 L/ha, 270–750 g/acre	No data	Low-input annual grass. Ethanol production possible using existing technology. Grows in tropical and temperate climates, but highest ethanol yield estimates assume multiple crops per year (possible only in tropical climates). Does not store well.
Corn	3100–4000 L/ha, 330–424 g/acre	10%–20%	High-input annual grass. Used as feedstock for most bioethanol produced in USA. Only kernels can be processed using available technology; development of commercial cellulosic technology would allow stover to be used and increase ethanol yield by 1,100 – 2,000 litres/ha.
Source (except those indicated): *Nature* 444 (7 December 2006): 673–676. [a] – Savings of GHG emissions assuming no land use change (using existing crop lands).			

Reduced Petroleum Imports and Costs

One rationale given for extensive ethanol production in the U.S. is its benefit to energy security, by shifting the need for some foreign-produced oil to domestically produced energy sources. Production of ethanol requires significant energy, but current U.S. production derives most of that energy from coal, natural gas and other sources, rather than oil. Because 66% of oil consumed in the U.S. is imported, compared to a net surplus of coal and just 16% of natural gas (figures from 2006), the displacement of oil-based fuels to ethanol produces a net shift from foreign to domestic U.S. energy sources.

According to a 2008 analysis by Iowa State University, the growth in US ethanol production has caused retail gasoline prices to be US $0.29 to US $0.40 per gallon lower than would otherwise have been the case.

Motorsport

Leon Duray qualified third for the 1927 Indianapolis 500 auto race with an ethanol-fueled car. The IndyCar Series adopted a 10% ethanol blend for the 2006 season, and a 98% blend in 2007.

In drag racing, there are Top Alcohol classes for dragsters and funny cars since the 1970s.

The American Le Mans Series sports car championship introduced E10 in the 2007 season to replace pure gasoline. In the 2008 season, E85 was allowed in the GT class and teams began switching to it.

In 2011, the three national NASCAR stock car series mandated a switch from gasoline to E15, a blend of Sunoco GTX unleaded racing fuel and 15% ethanol.

Australia's V8 Supercar championship uses United E85 for its racing fuel.

Stock Car Brasil Championship runs on neat ethanol, E100.

Ethanol fuel may also be utilized as a rocket fuel. As of 2010, small quantities of ethanol are used in lightweight rocket-racing aircraft.

Replacement of Kerosene for Lighting and Cooking

There is still extensive use of kerosene for lighting and cooking in less developed countries, and ethanol can have a role in reducing petroleum dependency in this use too. A non-profit named Project Gaia seeks to spread the use of ethanol stoves to replace wood, charcoal and kerosene. There is also potential for bioethanol replacing some kerosene use in domestic lighting from feedstocks grown locally.

Research

Ethanol plant in Turner County, South Dakota

Ethanol research focuses on alternative sources, novel catalysts and production processes.

In 2013, INEOS began initial operation of a bio-ethanol plant from vegetative material and wood waste.

The bacterium E.coli when genetically engineered with cow rumen genes and enzymes can produce ethanol from corn stover.

An alternative technology allows for the production of biodiesel from grain that has already been used to produce ethanol.

Another approach uses feed stocks such as municipal waste, recycled products, rice hulls, sugarcane bagasse, wood chips or switchgrass.

Algae Fuel

A conical flask of "green" jet fuel made from algae

Algae fuel or algal biofuel is an alternative to liquid fossil fuels that uses algae as its source of energy-rich oils. Also, algae fuels are an alternative to common known biofuel sources, such as corn and sugarcane. Several companies and government agencies are funding efforts to reduce capital and operating costs and make algae fuel production commercially viable. Like fossil fuel, algae fuel releases CO_2 when burnt, but unlike fossil fuel, algae fuel and other biofuels only release CO_2 recently removed from the atmosphere via photosynthesis as the algae or plant grew. The energy crisis and the world food crisis have ignited interest in algaculture (farming algae) for making biodiesel and other biofuels using land unsuitable for agriculture. Among algal fuels' attractive characteristics are that they can be grown with minimal impact on fresh water resources, can be produced using saline and wastewater, have a high flash point, and are biodegradable and relatively harmless to the environment if spilled. Algae cost more per unit mass than other second-generation biofuel crops due to high capital and operating costs, but are claimed to yield between 10 and 100 times more fuel per unit area. The United States Department of Energy estimates that if algae fuel replaced all the petroleum fuel in the United States, it would require 15,000 square miles (39,000 km²), which is only 0.42% of the U.S. map, or about half of the land area of Maine. This is less than $\frac{1}{7}$ the area of corn harvested in the United States in 2000.

According to the head of the Algal Biomass Organization, algae fuel can reach price parity with oil in 2018 if granted production tax credits. However, in 2013, Exxon Mobil Chairman and CEO Rex Tillerson said that after committing to spend up to $600 million over 10 years on development in a joint venture with J. Craig Venter's Synthetic Genomics in 2009, Exxon pulled back after four years (and $100 million) when it realized that algae fuel is "probably further" than 25 years away from commercial viability. On the other hand, Solazyme and Sapphire Energy already began commercial sales of algal biofuel in 2012 and 2013, respectively, and Algenol hopes to produce commercially in 2014.

History

In 1942 Harder and Von Witsch were the first to propose that microalgae be grown as a source of lipids for food or fuel. Following World War II, research began in the US, Germany, Japan, England, and Israel on culturing techniques and engineering systems for growing microalgae on larger scales, particularly species in the genus *Chlorella*. Meanwhile, H. G. Aach showed that *Chlorella pyrenoidosa* could be induced via nitrogen starvation to accumulate as much as 70% of its dry weight as lipids. Since the need for alternative transportation fuel had subsided after World War II, research at this time focused on culturing algae as a food source or, in some cases, for wastewater treatment.

Interest in the application of algae for biofuels was rekindled during the oil embargo and oil price surges of the 1970s, leading the US Department of Energy to initiate the Aquatic Species Program in 1978. The Aquatic Species Program spent $25 million over 18 years with the goal of developing liquid transportation fuel from algae that would be price competitive with petroleum-derived fuels. The research program focused on the cultivation of microalgae in open outdoor ponds, systems which are low in cost but vulnerable to environmental disturbances like temperature swings and biological invasions. 3,000 algal strains were collected from around the country and screened for desirable properties such as high productivity, lipid content, and thermal tolerance, and the most promising strains were included in the SERI microalgae collection at the Solar Energy Research Institute (SERI) in Golden, Colorado and used for further research. Among the program's most significant findings were that rapid growth and high lipid production were "mutually exclusive," since the former required high nutrients and the latter required low nutrients. The final report suggested that genetic engineering may be necessary to be able to overcome this and other natural limitations of algal strains, and that the ideal species might vary with place and season. Although it was successfully demonstrated that large-scale production of algae for fuel in outdoor ponds was feasible, the program failed to do so at a cost that would be competitive with petroleum, especially as oil prices sank in the 1990s. Even in the best case scenario, it was estimated that unextracted algal oil would cost $59–186 per barrel, while petroleum cost less than $20 per barrel in 1995. Therefore, under budget pressure in 1996, the Aquatic Species Program was abandoned.

Other contributions to algal biofuels research have come indirectly from projects focusing on different applications of algal cultures. For example, in the 1990s Japan's Research Institute of Innovative Technology for the Earth (RITE) implemented a research program with the goal of developing systems to fix CO_2 using microalgae. Although the goal was not energy production, several studies produced by RITE demonstrated that algae could be grown using flue gas from power plants as a CO_2 source, an important development for algal biofuel research. Other work focusing on harvesting hydrogen gas, methane, or ethanol from algae, as well as nutritional supplements and pharmaceutical compounds, has also helped inform research on biofuel production from algae.

Following the disbanding of the Aquatic Species Program in 1996, there was a relative lull in algal biofuel research. Still, various projects were funded in the US by the Department of Energy, Department of Defense, National Science Foundation, Department of Agriculture, National Laboratories, state funding, and private funding, as well as in other countries. More recently, rising oil prices in the 2000s spurred a revival of interest in algal biofuels and US federal funding has increased, numerous research projects are being funded in Australia, New Zealand, Europe, the Middle East, and other parts of the world, and a wave of private companies has entered the field. In November 2012, Solazyme and Propel Fuels made the first retail sales of algae-derived fuel, and in March 2013 Sapphire Energy began commercial sales of algal biofuel to Tesoro.

Fuels

Algae can be converted into various types of fuels, depending on the technique and the part of the cells used. The lipid, or oily part of the algae biomass can be extracted and converted into biodiesel through a process similar to that used for any other vegetable oil, or converted in a refinery into "drop-in" replacements for petroleum-based fuels. Alternatively or following lipid extraction, the carbohydrate content of algae can be fermented into bioethanol or butanol fuel.

Biodiesel

Biodiesel is a diesel fuel derived from animal or plant lipids (oils and fats). Studies have shown that some species of algae can produce 60% or more of their dry weight in the form of oil. Because the cells grow in aqueous suspension, where they have more efficient access to water, CO_2 and dissolved nutrients, microalgae are capable of producing large amounts of biomass and usable oil in either high rate algal ponds or photobioreactors. This oil can then be turned into biodiesel which could be sold for use in automobiles. Regional production of microalgae and processing into biofuels will provide economic benefits to rural communities.

As they do not have to produce structural compounds such as cellulose for leaves, stems, or roots, and because they can be grown floating in a rich nutritional medium, microalgae can have faster growth rates than terrestrial crops. Also, they can convert a much higher fraction of their biomass to oil than conventional crops, e.g. 60% versus 2-3% for soybeans. The per unit area yield of oil from algae is estimated to be from 58,700 to 136,900 L/ha/year, depending on lipid content, which is 10 to 23 times as high as the next highest yielding crop, oil palm, at 5,950 L/ha/year.

The U.S. Department of Energy's Aquatic Species Program, 1978–1996, focused on biodiesel from microalgae. The final report suggested that biodiesel could be the only viable method by which to produce enough fuel to replace current world diesel usage. If algae-derived biodiesel were to replace the annual global production of 1.1bn tons of conventional diesel then a land mass of 57.3 million hectares would be required, which would be highly favorable compared to other biofuels.

Biobutanol

Butanol can be made from algae or diatoms using only a solar powered biorefinery. This fuel has an energy density 10% less than gasoline, and greater than that of either ethanol or methanol. In most gasoline engines, butanol can be used in place of gasoline with no modifications. In several tests, butanol consumption is similar to that of gasoline, and when blended with gasoline, provides better performance and corrosion resistance than that of ethanol or E85.

The green waste left over from the algae oil extraction can be used to produce butanol. In addition, it has been shown that macroalgae (seaweeds) can be fermented by *Clostridia* genus bacteria to butanol and other solvents.

Biogasoline

Biogasoline is gasoline produced from biomass. Like traditionally produced gasoline, it contains between 6 (hexane) and 12 (dodecane) carbon atoms per molecule and can be used in internal-combustion engines.

Methane

Methane, the main constituent of natural gas can be produced from algae in various methods, namely Gasification, Pyrolysis and Anaerobic Digestion. In Gasification and Pyrolysis methods methane is extracted under high temperature and pressure. Anaerobic Digestion is a straight for-ward method involved in decomposition of algae into simple components then transforming it into fatty acids using microbes like acidific bacteria followed by removing any solid particles and finally adding methanogenic bacteria to release a gas mixture containing methane. A number of studies have successfully shown that biomass from microalgae can be converted into biogas via anaero-bic digestion. Therefore, in order to improve the overall energy balance of microalgae cultivation operations, it has been proposed to recover the energy contained in waste biomass via anaerobic digestion to methane for generating electricity.

Ethanol

The Algenol system which is being commercialized by BioFields in Puerto Libertad, Sonora, Mex-ico utilizes seawater and industrial exhaust to produce ethanol. Porphyridium cruentum also have shown to be potentially suitable for ethanol production due to its capacity for accumulating large amount of carbohydrates.

Hydrotreating to Traditional Transport Fuels

Algae can be used to produce 'green diesel' (also known as renewable diesel, hydrotreating veg-etable oil or hydrogen-derived renewable diesel) through a hydrotreating refinery process that breaks molecules down into shorter hydrocarbon chains used in diesel engines. It has the same chemical properties as petroleum-based diesel meaning that it does not require new engines, pipe-lines or infrastructure to distribute and use. It has yet to be produced at a cost that is competitive with petroleum. While hydrotreating is currently the most common pathway to produce fuel-like hydrocarbons via decarboxylation/decarbonylation, an alternative process offering a number of important advantages over hydrotreating. In this regard, the work of Crocker et al. and Lercher et al. is particularly noteworthy. for of oil refining, research is underway for catalytic conversion of renewable fuels by decarboxylation.

Jet Fuel

Rising jet fuel prices are putting severe pressure on airline companies, creating an incentive for al-gal jet fuel research. The International Air Transport Association, for example, supports research, development and deployment of algal fuels. IATA's goal is for its members to be using 10% alternative fuels by 2017.

Trials have been carried with aviation biofuel by Air New Zealand, Lufthansa, and Virgin Airlines.

In February 2010, the Defense Advanced Research Projects Agency announced that the U.S. military was about to begin large-scale oil production from algal ponds into jet fuel. After extraction at a cost of $2 per gallon, the oil will be refined at less than $3 a gallon. A larger-scale refining operation, producing 50 million gallons a year, is expected to go into production in 2013, with the possibility of lower per gallon costs so that algae-based fuel would be competitive with fossil fuels. The projects, run by the companies SAIC and General Atomics, are expected to produce 1,000 gallons of oil per acre per year from algal ponds.

Species

Research into algae for the mass-production of oil focuses mainly on microalgae (organisms capable of photosynthesis that are less than 0.4 mm in diameter, including the diatoms and cyanobacteria) as opposed to macroalgae, such as seaweed. The preference for microalgae has come about due largely to their less complex structure, fast growth rates, and high oil-content (for some species). However, some research is being done into using seaweeds for biofuels, probably due to the high availability of this resource.

As of 2012 researchers across various locations worldwide have started investigating the following species for their suitability as a mass oil-producers:

- *Botryococcus braunii*

- *Chlorella*

- *Dunaliella tertiolecta*

- *Gracilaria*

- *Pleurochrysis carterae* (also called CCMP647).

- *Sargassum*, with 10 times the output volume of *Gracilaria*.

The amount of oil each strain of algae produces varies widely. Note the following microalgae and their various oil yields:

- *Ankistrodesmus* TR-87: 28–40% dry weight

- *Botryococcus braunii*: 29–75% dw

- *Chlorella* sp.: 29%dw

- *Chlorella protothecoides*(autotrophic/ heterothrophic): 15–55% dw

- *Crypthecodinium cohnii*: 20%dw

- *Cyclotella* DI- 35: 42%dw

- *Dunaliella tertiolecta* : 36–42%dw

- *Hantzschia* DI-160: 66%dw

- *Nannochloris*: 31(6–63)%dw

- *Nannochloropsis* : 46(31–68)%dw

- *Neochloris oleoabundans*: 35–54%dw

- *Nitzschia* TR-114: 28–50%dw

- *Phaeodactylum tricornutum*: 31%dw

- *Scenedesmus* TR-84: 45%dw

- *Schizochytrium* 50–77%dw

- *Stichococcus*: 33(9–59)%dw

- *Tetraselmis suecica*: 15–32%dw

- *Thalassiosira pseudonana*: (21–31)%dw

In addition, due to its high growth-rate, *Ulva* has been investigated as a fuel for use in the *SOFT cycle*, (SOFT stands for Solar Oxygen Fuel Turbine), a closed-cycle power-generation system suitable for use in arid, subtropical regions.

Algae Cultivation

Photobioreactor from glass tubes

Open Pond

Design of a race-way open pond commonly used for algal culture

Algae grow much faster than food crops, and can produce hundreds of times more oil per unit area than conventional crops such as rapeseed, palms, soybeans, or jatropha. As algae have a harvesting cycle of 1–10 days, their cultivation permits several harvests in a very short time-frame, a strategy differing from that associated with annual crops. In addition, algae can be grown on land unsuitable for terrestrial crops, including arid land and land with excessively saline soil, minimizing competition with agriculture. Most research on algae cultivation has focused on growing algae in clean but expensive photobioreactors, or in open ponds, which are cheap to maintain but prone to contamination.

Closed-loop System

The lack of equipment and structures needed to begin growing algae in large quantities has inhibited widespread mass-production of algae for biofuel production. Maximum use of existing agriculture processes and hardware is the goal.

Closed systems (not exposed to open air) avoid the problem of contamination by other organisms blown in by the air. The problem for a closed system is finding a cheap source of sterile CO_2. Several experimenters have found the CO_2 from a smokestack works well for growing algae. For reasons of economy, some experts think that algae farming for biofuels will have to be done as part of cogeneration, where it can make use of waste heat and help soak up pollution.

Photobioreactors

Most companies pursuing algae as a source of biofuels pump nutrient-rich water through plastic or borosilicate glass tubes (called "bioreactors") that are exposed to sunlight (and so-called photobioreactors or PBR).

Running a PBR is more difficult than using an open pond, and costlier, but may provide a higher level of control and productivity. In addition, a photobioreactor can be integrated into a closed loop cogeneration system much more easily than ponds or other methods.

Open Pond

Raceway pond used for the cultivation of microalgae

Open-pond systems for the most part have been given up for the cultivation of algae with especially high oil content. Many believe that a major flaw of the Aquatic Species Program was the decision to focus their efforts exclusively on open-ponds; this makes the entire effort dependent upon the hardiness of the strain chosen, requiring it to be unnecessarily resilient in order to withstand wide swings in temperature and pH, and competition from invasive algae and bacteria. Open systems using a monoculture are also vulnerable to viral infection. The energy that a high-oil strain invests into the production of oil is energy that is not invested into the production of proteins or carbohydrates, usually resulting in the species being less hardy, or having a slower growth rate. Algal species with a lower oil content, not having to divert their energies away from growth, can be grown more effectively in the harsher conditions of an open system.

Some open sewage-ponds trial production has taken place in Marlborough, New Zealand.

Algal Turf Scrubber

2.5 acre ATS system, installed by Hydromentia on a farm creek in Florida

The algal turf scrubber (ATS) is a system designed primarily for cleaning nutrients and pollutants out of water using algal turfs. ATS mimics the algal turfs of a natural coral reef by taking in nutrient rich water from waste streams or natural water sources, and pulsing it over a sloped surface. This surface is coated with a rough plastic membrane or a screen, which allows naturally occurring algal spores to settle and colonize the surface. Once the algae has been established, it can be harvested every 5–15 days, and can produce 18 metric tons of algal biomass per hectare per year. In contrast to other methods, which focus primarily on a single high yielding species of algae, this method focuses on naturally occurring polycultures of algae. As such, the lipid content of the algae in an ATS system is usually lower, which makes it more suitable for a fermented fuel product, such as ethanol, methane, or butanol. Conversely, the harvested algae could be treated with a hydrothermal liquefaction process, which would make possible biodiesel, gasoline, and jet fuel production.

There are three major advantages of ATS over other systems. The first advantage is documented higher productivity over open pond systems. The second is lower operating and fuel production costs. The third is the elimination of contamination issues due to the reliance on naturally occurring algae species. The projected costs for energy production in an ATS system are $0.75/kg, compared to a photobioreactor which would cost $3.50/kg. Furthermore, due to the fact that the primary purpose of ATS is removing nutrients and pollutants out of water, and these costs have been shown to be lower than other methods of nutrient removal, this may incentivize the use of this technology for nutrient removal as the primary function, with biofuel production as an added benefit.

Algae being harvested and dried from an ATS system

Fuel Production

After harvesting the algae, the biomass is typically processed in a series of steps, which can differ based on the species and desired product; this is an active area of research. and also is the bottleneck of this technology: the cost of extraction is higher than those obtained. One of the solutions is to use filter feeders to "eat" them. Improved animals can provide both foods and fuels.

Dehydration

Often, the algae is dehydrated, and then a solvent such as hexane is used to extract energy-rich compounds like triglycerides from the dried material. Then, the extracted compounds can be processed into fuel using standard industrial procedures. For example, the extracted triglycerides are reacted with methanol to create biodiesel via transesterification. The unique composition of fatty acids of each species influences the quality of the resulting biodiesel and thus must be taken into account when selecting algal species for feedstock.

Hydrothermal Liquefaction

An alternative approach called Hydrothermal liquefaction employs a continuous process that subjects harvested wet algae to high temperatures and pressures—350 °C (662 °F) and 3,000 pounds per square inch (21,000 kPa).

Products include crude oil, which can be further refined into aviation fuel, gasoline, or diesel fuel using one or many upgrading processes. The test process converted between 50 and 70 percent of the algae's carbon into fuel. Other outputs include clean water, fuel gas and nutrients such as nitrogen, phosphorus, and potassium.

Nutrients

Nutrients like nitrogen (N), phosphorus (P), and potassium (K), are important for plant growth and are essential parts of fertilizer. Silica and iron, as well as several trace elements, may also be considered important marine nutrients as the lack of one can limit the growth of, or productivity in, an area.

Carbon Dioxide

Bubbling CO_2 through algal cultivation systems can greatly increase productivity and yield (up to a saturation point). Typically, about 1.8 tonnes of CO_2 will be utilised per tonne of algal biomass (dry) produced, though this varies with algae species. The Glenturret Distillery in Perthshire, UK – home to The Famous Grouse Whisky – percolate CO_2 made during the whisky distillation through a microalgae bioreactor. Each tonne of microalgae absorbs two tonnes of CO_2. Scottish Bioenergy, who run the project, sell the microalgae as high value, protein-rich food for fisheries. In the future, they will use the algae residues to produce renewable energy through anaerobic digestion.

Nitrogen

Nitrogen is a valuable substrate that can be utilized in algal growth. Various sources of nitrogen can be used as a nutrient for algae, with varying capacities. Nitrate was found to be the preferred source of nitrogen, in regards to amount of biomass grown. Urea is a readily available source that shows comparable results, making it an economical substitute for nitrogen source in large scale culturing of algae. Despite the clear increase in growth in comparison to a nitrogen-less medium, it has been shown that alterations in nitrogen levels affect lipid content within the algal cells. In one study nitrogen deprivation for 72 hours caused the total fatty acid content (on a per cell basis) to increase by 2.4-fold. 65% of the total fatty acids were esterified to triacylglycerides in oil bodies, when compared to the initial culture, indicating that the algal cells utilized de novo synthesis of fatty acids. It is vital for the lipid content in algal cells to be of high enough quantity, while maintaining adequate cell division times, so parameters that can maximize both are under investigation.

Wastewater

A possible nutrient source is waste water from the treatment of sewage, agricultural, or flood plain run-off, all currently major pollutants and health risks. However, this waste water cannot feed algae directly and must first be processed by bacteria, through anaerobic digestion. If waste water is not processed before it reaches the algae, it will contaminate the algae in the reactor, and at the very least, kill much of the desired algae strain. In biogas facilities, organic waste is often converted to a mixture of carbon dioxide, methane, and organic fertilizer. Organic fertilizer that comes out of the digester is liquid, and nearly suitable for algae growth, but it must first be cleaned and sterilized.

The utilization of wastewater and ocean water instead of freshwater is strongly advocated due to the continuing depletion of freshwater resources. However, heavy metals, trace metals, and other contaminants in wastewater can decrease the ability of cells to produce lipids biosynthetically and also impact various other workings in the machinery of cells. The same is true for ocean water, but the contaminants are found in different concentrations. Thus, agricultural-grade fertilizer is the preferred source of nutrients, but heavy metals are again a problem, especially for strains of algae that are susceptible to these metals. In open pond systems the use of strains of algae that can deal with high concentrations of heavy metals could prevent other organisms from infesting these systems. In some instances it has even been shown that strains of algae can remove over 90% of nickel and zinc from industrial wastewater in relatively short periods of time.

Environmental Impact

In comparison with terrestrial-based biofuel crops such as corn or soybeans, microalgal production results in a much less significant land footprint due to the higher oil productivity from the microalgae than all other oil crops. Algae can also be grown on marginal lands useless for ordinary crops and with low conservation value, and can use water from salt aquifers that is not useful for agriculture or drinking., Algae can also grow on the surface of the ocean in bags or floating screens. Thus microalgae could provide a source of clean energy with little impact on the provisioning of adequate food and water or the conservation of biodiversity. Algae cultivation also requires no external subsidies of insecticides or herbicides, removing any risk of generating associated pesticide waste streams. In addition, algal biofuels are much less toxic, and degrade far more readily than petroleum-based fuels. However, due to the flammable nature of any combustible fuel, there is potential for some environmental hazards if ignited or spilled, as may occur in a train derailment or a pipeline leak. This hazard is reduced compared to fossil fuels, due to the ability for algal biofuels to be produced in a much more localized manner, and due to the lower toxicity overall, but the hazard is still there nonetheless. Therefore, algal biofuels should be treated in a similar manner to petroleum fuels in transportation and use, with sufficient safety measures in place at all times.

Studies have determined that replacing fossil fuels with renewable energy sources, such as biofuels, have the capability of reducing CO_2 emissions by up to 80%. An algae-based system could capture approximately 80% of the CO_2 emitted from a power plant when sunlight is available. Although this CO_2 will later be released into the atmosphere when the fuel is burned, this CO_2 would have entered the atmosphere regardless. The possibility of reducing total CO_2 emissions therefore lies in the prevention of the release of CO_2 from fossil fuels. Furthermore, compared to fuels like diesel and petroleum, and even compared to other sources of biofuels, the production and combustion of algal biofuel does not produce any sulfur oxides or nitrous oxides, and produces a reduced amount of carbon monoxide, unburned hydrocarbons, and reduced emission of other harmful pollutants. Since terrestrial plant sources of biofuel production simply do not have the production capacity to meet current energy requirements, microalgae may be one of the only options to approach complete replacement of fossil fuels.

Microalgae production also includes the ability to use saline waste or waste CO_2 streams as an energy source. This opens a new strategy to produce biofuel in conjunction with waste water treatment, while being able to produce clean water as a byproduct. When used in a microalgal bioreactor, harvested microalgae will capture significant quantities of organic compounds as well as heavy metal contaminants absorbed from wastewater streams that would otherwise be directly discharged into surface and ground-water. Moreover, this process also allows the recovery of phosphorus from waste, which is an essential but scarce element in nature – the reserves of which are estimated to have depleted in the last 50 years. Another possibility is the use of algae production systems to clean up non-point source pollution, in a system known as an algal turf scrubber (ATS). This has been demonstrated to reduce nitrogen and phosphorus levels in rivers and other large bodies of water affected by eutrophication, and systems are being built that will be capable of processing up to 110 million liters of water per day. ATS can also be used for treating point source pollution, such as the waste water mentioned above, or in treating livestock effluent.

Polycultures

Nearly all research in algal biofuels has focused on culturing single species, or monocultures, of microalgae. However, ecological theory and empirical studies have demonstrated that plant and algae polycultures, i.e. groups of multiple species, tend to produce larger yields than monocultures. Experiments have also shown that more diverse aquatic microbial communities tend to be more stable through time than less diverse communities. Recent studies found that polycultures of microalgae produced significantly higher lipid yields than monocultures. Polycultures also tend to be more resistant to pest and disease outbreaks, as well as invasion by other plants or algae. Thus culturing microalgae in polyculture may not only increase yields and stability of yields of biofuel, but also reduce the environmental impact of an algal biofuel industry.

Economic Viability

There is clearly a demand for sustainable biofuel production, but whether a particular biofuel will be used ultimately depends not on sustainability but cost efficiency. Therefore, research is focusing on cutting the cost of algal biofuel production to the point where it can compete with conventional petroleum. The production of several products from algae has been mentioned as the most important factor for making algae production economically viable. Other factors are the improving of the solar energy to biomass conversion efficiency (currently 3%, but 5 to 7% is theoretically attainable) and making the oil extraction from the algae easier.

In a 2007 report a formula was derived estimating the cost of algal oil in order for it to be a viable substitute to petroleum diesel:

$$C_{(algal\ oil)} = 25.9 \times 10^{-3}\ C_{(petroleum)}$$

where: $C_{(algal\ oil)}$ is the price of microalgal oil in dollars per gallon and $C_{(petroleum)}$ is the price of crude oil in dollars per barrel. This equation assumes that algal oil has roughly 80% of the caloric energy value of crude petroleum.

With current technology available, it is estimated that the cost of producing microalgal biomass is $2.95/kg for photobioreactors and $3.80/kg for open-ponds. These estimates assume that carbon dioxide is available at no cost. If the annual biomass production capacity is increased to 10,000 tonnes, the cost of production per kilogram reduces to roughly $0.47 and $0.60, respectively. Assuming that the biomass contains 30% oil by weight, the cost of biomass for providing a liter of oil would be approximately $1.40 ($5.30/gal) and $1.81 ($6.85/gal) for photobioreactors and raceways, respectively. Oil recovered from the lower cost biomass produced in photobioreactors is estimated to cost $2.80/L, assuming the recovery process contributes 50% to the cost of the final recovered oil. If existing algae projects can achieve biodiesel production price targets of less than $1 per gallon, the United States may realize its goal of replacing up to 20% of transport fuels by 2020 by using environmentally and economically sustainable fuels from algae production.

Whereas technical problems, such as harvesting, are being addressed successfully by the industry, the high up-front investment of algae-to-biofuels facilities is seen by many as a major obstacle to the success of this technology. Only few studies on the economic viability are publicly available, and must often rely on the little data (often only engineering estimates) available in the public domain. Dmitrov examined the GreenFuel's photobioreactor and estimated that algae oil would only be competitive at an oil price of $800 per barrel. A study by Alabi et al. examined raceways, photobioreactors and anaerobic fermenters to make biofuels from algae and found that photobioreactors are too expensive to make biofuels. Raceways might be cost-effective in warm climates with very low labor costs, and fermenters may become cost-effective subsequent to significant process improvements. The group found that capital cost, labor cost and operational costs (fertilizer, electricity, etc.) by themselves are too high for algae biofuels to be cost-competitive with conventional fuels. Similar results were found by others, suggesting that unless new, cheaper ways of harnessing algae for biofuels production are found, their great technical potential may never become economically accessible. Recently, Rodrigo E. Teixeira demonstrated a new reaction and proposed a process for harvesting and extracting raw materials for biofuel and chemical production that requires a fraction of the energy of current methods, while extracting all cell constituents.

Use of Byproducts

Many of the byproducts produced in the processing of microalgae can be used in various applications, many of which have a longer history of production than algal biofuel. Some of the products not used in the production of biofuel include natural dyes and pigments, antioxidants, and other high-value bio-active compounds. These chemicals and excess biomass have found numerous use in other industries. For example, the dyes and oils have found a place in cosmetics, commonly as thickening and water-binding agents. Discoveries within the pharmaceutical industry include antibiotics and antifungals derived from microalgae, as well as natural health products, which have been growing in popularity over the past few decades. For instance *Spirulina* contains numerous polyunsaturated fats (Omega 3 and 6), amino acids, and vitamins, as well as pigments that may be beneficial, such as beta-carotene and chlorophyll.

Advantages

Ease of Growth

One of the main advantages that using microalgae as the feedstock when compared to more traditional crops is that it can be grown much more easily. Algae can be grown in land that would not be considered suitable for the growth of the regularly used crops. In addition to this, wastewater that would normally hinder plant growth has been shown to be very effective in growing algae. Because of this, algae can be grown without taking up arable land that would otherwise be used for producing food crops, and the better resources can be reserved for normal crop production. Microalgae also require fewer resources to grow and little attention is needed, allowing the growth and cultivation of algae to be a very passive process.

Impact on Food

Many traditional feedstocks for biodiesel, such as corn and palm, are also used as feed for livestock on farms, as well as a valuable source of food for humans. Because of this, using them as biofuel reduces the amount of food available for both, resulting in an increased cost for both the food and the fuel produced. Using algae as a source of biodiesel can alleviate this problem in a number of ways. First, algae is not used as a primary food source for humans, meaning that it can be used solely for fuel and there would be little impact in the food industry. Second, many of the waste-product extracts produced during the processing of algae for biofuel can be used as a sufficient animal feed. This is an effective way to minimize waste and a much cheaper alternative to the more traditional corn- or grain-based feeds.

Minimization of Waste

Growing algae as a source of biofuel has also been shown to have numerous environmental benefits, and has presented itself as a much more environmentally friendly alternative to current biofuels. For one, it is able to utilize run-off, water contaminated with fertilizers and other nutrients that are a by-product of farming, as its primary source of water and nutrients. Because of this, it prevents this contaminated water from mixing with the lakes and rivers that currently supply our drinking water. In addition to this, the ammonia, nitrates, and phosphates that would normally render the water unsafe actually serve as excellent nutrients for the algae, meaning that fewer resources are needed to grow the algae. Many algae species used in biodiesel production are excellent bio-fixers, meaning they are able to remove carbon dioxide from the atmosphere to use as a form of energy for themselves. Because of this, they have found use in industry as a way to treat flue gases and reduce GHG emissions.

Disadvantages

Commercial Viability

Algae biodiesel is still a fairly new technology. Despite the fact that research began over 30 years ago, it was put on hold during the mid-1990s, mainly due to a lack of funding and a relatively low petroleum cost. For the next few years algae biofuels saw little attention; it was not until the gas peak of the early 2000s that it eventually had a revitalization in the search for alternative fuel sources. While the technology exists to harvest and convert algae into a usable source of biodiesel, it still hasn't been implemented into a large enough scale to support the current energy needs. Further research will be required to make the production of algae biofuels more efficient, and at this point it is currently being held back by lobbyists in support of alternative biofuels, like those produced from corn and grain. In 2013, Exxon Mobil Chairman and CEO Rex Tillerson said that after originally committing to spending up to $600 million on development in a joint venture with J. Craig Venter's Synthetic Genomics, algae is "probably further" than "25 years away" from commercial viability, although Solazyme and Sapphire Energy already began small-scale commercial sales in 2012 and 2013, respectively.

Stability

The biodiesel produced from the processing of microalgae differs from other forms of biodiesel in the content of polyunsaturated fats. Polyunsaturated fats are known for their ability to retain fluidity at lower temperatures. While this may seem like an advantage in production during the colder temperatures of the winter, the polyunsaturated fats result in lower stability during regular seasonal temperatures.

Research

United States

The National Renewable Energy Laboratory (NREL) is the U.S. Department of Energy's primary national laboratory for renewable energy and energy efficiency research and development. This program is involved in the production of renewable energies and energy efficiency. One of its most current divisions are consists the biomass program which is involved in biomass characterization, biochemical and thermochemical conversion technologies in conjunction with biomass process engineering and analysis. The program aims at producing energy efficient, cost-effective and environmentally friendly technologies that support rural economies, reduce the nations dependency in oil and improve air quality.

At the Woods Hole Oceanographic Institution and the Harbor Branch Oceanographic Institution the wastewater from domestic and industrial sources contain rich organic compounds that are being used to accelerate the growth of algae. The Department of Biological and Agricultural Engineering at University of Georgia is exploring microalgal biomass production using industrial wastewater. Algaewheel, based in Indianapolis, Indiana, presented a proposal to build a facility in Cedar Lake, Indiana that uses algae to treat municipal wastewater, using the sludge byproduct to produce biofuel. A similar approach is being followed by Algae Systems, a company based in Daphne, Alabama.

Sapphire Energy (San Diego) has produced green crude from algae.

Solazyme (South San Francisco, California) has produced a fuel suitable for powering jet aircraft from algae.

Europe

Universities in the United Kingdom which are working on producing oil from algae include: University of Manchester, University of Sheffield, University of Glasgow, University of Brighton, University of Cambridge, University College London, Imperial College London, Cranfield University and Newcastle University. In Spain, it is also relevant the research carried out by the CSIC´s Instituto de Bioquímica Vegetal y Fotosíntesis (Microalgae Biotechnology Group, Seville).

The Marine Research station in Ketch Harbour, Nova Scotia, has been involved in growing algae for 50 years. The National Research Council (Canada) (NRC) and National Byproducts Program have provided $5 million to fund this project. The aim of the program has been to build a 50 000 litre cultivation pilot plant at the Ketch harbor facility. The station has been involved in assessing how best to grow algae for biofuel and is involved in investigating the utilization of numerous algae species in regions of North America. NRC has joined forces with the United States Department of Energy, the National Renewable Energy Laboratory in Colorado and Sandia National Laboratories in New Mexico.

The European Algae Biomass Association (EABA) is the European association representing both research and industry in the field of algae technologies, currently with 79 members. The association is headquartered in Florence, Italy. The general objective of the EABA is to promote mutual interchange and cooperation in the field of biomass production and use, including biofuels uses and all other utilisations. It aims at creating, developing and maintaining solidarity and links between its Members and at defending their interests at European and international level. Its main target is to act as a catalyst for fostering synergies among scientists, industrialists and decision makers to promote the development of research, technology and industrial capacities in the field of Algae.

CMCL innovations and the University of Cambridge are carrying out a detailed design study of a C-FAST (Carbon negative Fuels derived from Algal and Solar Technologies) plant. The main objective is to design a pilot plant which can demonstrate production of hydrocarbon fuels (including diesel and gasoline) as sustainable carbon-negative energy carriers and raw materials for the chemical commodity industry. This project will report in June 2013.

Ukraine plans to produce biofuel using a special type of algae.

The European Commission's Algae Cluster Project, funded through the Seventh Framework Programme, is made up of three algae biofuel projects, each looking to design and build a different algae biofuel facility covering 10ha of land. The projects are BIOFAT, All-Gas and InteSusAl.

Since various fuels and chemicals can be produced from algae, it has been suggested to investigate the feasibility of various production processes(conventional extraction/separation, hydrothermal liquefaction, gasification and pyrolysis) for application in an integrated algal biorefinery.

India

Reliance industries in collaboration with Algenol, USA commissioned a pilot project to produce algal bio-oil in the year 2014. Spirulina which is an alga rich in proteins content has been commercially cultivated in India. Algae is used in India for treating the sewage in open/natural oxidation ponds This reduces the Biological Oxygen Demand (BOD) of the sewage and also provides algal biomass which can be converted to fuel.

Other

The Algae Biomass Organization (ABO) is a non-profit organization whose mission is "to promote the development of viable commercial markets for renewable and sustainable commodities derived from algae".

The National Algae Association (NAA) is a non-profit organization of algae researchers, algae production companies and the investment community who share the goal of commercializing algae oil as an alternative feedstock for the biofuels markets. The NAA gives its members a forum to efficiently evaluate various algae technologies for potential early stage company opportunities.

Pond Biofuels Inc. in Ontario, Canada has a functioning pilot plant where algae is grown directly off of smokestack emissions from a cement plant, and dried using waste heat. In May 2013, Pond Biofuels announced a partnership with the National Research Council of Canada and Canadian Natural Resources Limited to construct a demonstration-scale algal biorefinery at an oil sands site near Bonnyville, Alberta.

Ocean Nutrition Canada in Halifax, Nova Scotia, Canada has found a new strain of algae that appears capable of producing oil at a rate 60 times greater than other types of algae being used for the generation of biofuels.

VG Energy, a subsidiary of Viral Genetics Incorporated, claims to have discovered a new method of increasing algal lipid production by disrupting the metabolic pathways that would otherwise divert photosynthetic energy towards carbohydrate production. Using these techniques, the company states that lipid production could be increased several-fold, potentially making algal biofuels cost-competitive with existing fossil fuels.

Algae production from the warm water discharge of a nuclear power plant has been piloted by Patrick C. Kangas at Peach Bottom Nuclear Power Station, owned by Exelon Corporation. This process takes advantage of the relatively high temperature water to sustain algae growth even during winter months.

Companies such as Sapphire Energy and Bio Solar Cells are using genetic engineering to make algae fuel production more efficient. According to Klein Lankhorst of Bio Solar Cells, genetic engineering could vastly improve algae fuel efficiency as algae can be modified to only build short carbon chains instead of long chains of carbohydrates. Sapphire Energy also uses chemically induced mutations to produce algae suitable for use as a crop.

Some commercial interests into large-scale algal-cultivation systems are looking to tie in to existing infrastructures, such as cement factories, coal power plants, or sewage treatment facilities. This approach changes wastes into resources to provide the raw materials, CO2 and nutrients, for the system.

A feasibility study using marine microalgae in a photobioreactor is being done by The International Research Consortium on Continental Margins at the Jacobs University Bremen.

The Department of Environmental Science at Ateneo de Manila University in the Philippines, is working on producing biofuel from a local species of algae.

Genetic Engineering

Genetic engineering algae has been used to increase lipid production or growth rates. Current research in genetic engineering includes either the introduction or removal of enzymes. In 2007 Oswald et al. introduced a monoterpene synthase from sweet basil into Saccharomyces cerevisiae, a strain of yeast. This particular monoterpene synthase causes the de novo synthesis of large amounts of geraniol, while also secreting it into the medium. Geraniol is a primary component in rose oil, palmarosa oil, and citronella oil as well as essential oils, making it a viable source of triacylglycerides for biodiesel production.

The enzyme ADP-glucose pyrophosphorylase is vital in starch production, but has no connection to lipid synthesis. Removal of this enzyme resulted in the sta6 mutant, which showed increased lipid content. After 18 hours of growth in nitrogen deficient medium the sta6 mutants had on average 17 ng triacylglycerides/1000 cells, compared to 10 ng/1000 cells in WT cells. This increase in lipid production was attributed to reallocation of intracellular resources, as the algae diverted energy from starch production.

In 2013 researchers used a "knock-down" of fat-reducing enzymes (multifunctional lipase/phospholipase/acyltransferase) to increase lipids (oils) without compromising growth. The study also introduced an efficient screening process. Antisense-expressing knockdown strains 1A6 and 1B1 contained 2.4- and 3.3-fold higher lipid content during exponential growth, and 4.1- and 3.2-fold higher lipid content after 40 h of silicon starvation.

Funding Programs

Numerous Funding programs have been created with aims of promoting the use of Renewable Energy. In Canada, the ecoAgriculture biofuels capital initiative (ecoABC) provides $25 million per project to assist farmers in constructing and expanding a renewable fuel production facility. The program has $186 million set aside for these projects. The sustainable development (SDTC) program has also applied $500 millions over 8 years to assist with the construction of next-generation renewable fuels. In addition, over the last 2 years $10 million has been made available for renewable fuel research and analysis

In Europe, the Seventh Framework Programme (FP7) is the main instrument for funding research. Similarly, the NER 300 is an unofficial, independent portal dedicated to renewable energy and grid integration projects. Another program includes the horizon 2020 program which will start 1 January, and will bring together the framework program and other EC innovation and research funding into a new integrated funding system

The American NBB's Feedstock Development program is addressing production of algae on the horizon to expand available material for biodiesel in a sustainable manner.

International Policies

Canada

Numerous policies have been put in place since the 1975 oil crisis in order to promote the use of Renewable Fuels in the United States, Canada and Europe. In Canada, these included the implementation of excise taxes exempting propane and natural gas which was extended to ethanol made from biomass and methanol in 1992. The federal government also announced their renewable fuels strategy in 2006 which proposed four components: increasing availability of renewable fuels through regulation, supporting the expansion of Canadian production of renewable fuels, assisting farmers to seize new opportunities in this sector and accelerating the commercialization of new technologies. These mandates were quickly followed by the Canadian provinces:

BC introduced a 5% ethanol and 5% renewable diesel requirement which was effective by January 2010. It also introduced a low carbon fuel requirement for 2012 to 2020.

Alberta introduced a 5% ethanol and 2% renewable diesel requirement implemented April 2011. The province also introduced a minimum 25% GHG emission reduction requirement for qualifying renewable fuels.

Saskatchewan implemented a 2% renewable diesel requirement in 2009.

Additionally, in 2006, the Canadian Federal Government announced its commitment to using its purchasing power to encourage the biofuel industry. Section three of the 2006 alternative fuels act stated that when it is economically feasible to do so-75% per cent of all federal bodies and crown corporation will be motor vehicles.

The National Research Council of Canada has established research on Algal Carbon Conversion as one of its flagship programs. As part of this program, the NRC made an announcement in May 2013 that they are partnering with Canadian Natural Resources Limited and Pond Biofuels to construct a demonstration-scale algal biorefinery near Bonnyville, Alberta.

United States

Policies in the United States have included a decrease in the subsidies provided by the federal and state governments to the oil industry which have usually included $2.84 billion. This is more than what is actually set aside for the biofuel industry. The measure was discussed at the G20 in Pittsburgh where leaders agreed that "inefficient fossil fuel subsidies encourage wasteful consumption, reduce our energy security, impede investment in clean sources and undermine efforts to deal with the threat of climate change". If this commitment is followed through and subsidies are removed, a fairer market in which algae biofuels can compete will be created. In 2010, the U.S. House of Representatives passed a legislation seeking to give algae-based biofuels parity with cellulose biofuels in federal tax credit programs. The algae-based renewable fuel promotion act (HR 4168) was implemented to give biofuel projects access to a $1.01 per gal production tax credit and 50% bonus depreciation for biofuel plant property. The U.S Government also introduced the domestic Fuel for Enhancing National Security Act implemented in 2011. This policy constitutes an amendment to the Federal property and administrative services act of 1949 and federal defense provisions in order to extend to 15 the number of years that the Department of Defense (DOD) multiyear contract may be entered into the case of the purchase of advanced biofuel. Federal and DOD programs are usually limited to a 5-year period

Other

The European Union (EU) has also responded by quadrupling the credits for second-generation algae biofuels which was established as an amendment to the Biofuels and Fuel Quality Directives

Companies

With algal biofuel being a relatively new alternative to conventional petroleum products, it leaves numerous opportunities for drastic advances in all aspects of the technology. Producing algae biofuel is not yet a cost-effective replacement for gasoline, but alterations to current methodologies can change this. The two most common targets for advancements are the growth medium (open pond vs. photobioreactor) and methods to remove the intracellular components of the algae. Below are companies that are currently innovating algal biofuel technologies.

Algenol Biofuels

Founded in 2006, Algenol Biofuels is a global, industrial biotechnology company that is commercializing its patented algae technology for production of ethanol and other fuels. Based in Southwest Florida, Algenol's patented technology enables the production of the four most important fuels (ethanol, gasoline, jet, and diesel fuel) using proprietary algae, sunlight, carbon dioxide and saltwater for around $1.27 per gallon and at production levels of 8,000 total gallons of liquid fuel per acre per year. Algenol's technology produces high yields and relies on patented photobioreactors and proprietary downstream techniques for low-cost fuel production using carbon dioxide from industrial sources.

Blue Marble Production

Blue Marble Production is a Seattle-based company that is dedicated to removing algae from algae-infested water. This in turn cleans up the environment and allows this company to produce biofuel. Rather than just focusing on the mass production of algae, this company focuses on what to do with the byproducts. This company recycles almost 100% of its water via reverse osmosis, saving about 26,000 gallons of water every month. This water is then pumped back into their system. The gas produced as a byproduct of algae will also be recycled by being placed into a photobioreactor system that holds multiple strains of algae. Whatever gas remains is then made into pyrolysis oil by thermochemical processes. Not only does this company seek to produce biofuel, but it also wishes to use algae for a variety of other purposes such as fertilizer, food flavoring, anti-inflammatory, and anti-cancer drugs.

Solazyme

Solazyme is one of a handful of companies which is supported by oil companies such as Chevron. Additionally, this company is also backed by Imperium Renewables, Blue Crest Capital Finance, and The Roda Group. Solazyme has developed a way to use up to 80% percent of dry algae as oil. This process requires the algae to grow in a dark fermentation vessel and be fed by carbon substrates within their growth media. The effect is the production of triglycerides that are almost identical to vegetable oil. Solazyme's production method is said to produce more oil than those algae cultivated photosynthetically or made to produce ethanol. Oil refineries can then take this algal oil and turn it into biodiesel, renewable diesel or jet fuels.

Part of Solazyme's testing, in collaboration with Maersk Line and the US Navy, placed 30 tons of Soladiesel(RD) algae fuel into the 98,000-tonne, 300-meter container ship Maersk Kalmar. This fuel was used at blends from 7% to 100% in an auxiliary engine on a month-long trip from Bremerhaven, Germany to Pipavav, India in Dec 2011. In Jul 2012, The US Navy used 700,000 gallons of HRD76 biodiesel in three ships of the USS Nimitz "Green Strike Group" during the 2012 RIMPAC exercise in Hawaii. The Nimitz also used 200,000 gallons of HRJ5 jet biofuel. The 50/50 biofuel blends were provided by Solazyme and Dynamic Fuels.

Sapphire Energy

Sapphire Energy is a leader in the algal biofuel industry backed by the Wellcome Trust, Bill Gates' Cascade Investment, Monsanto, and other large donors. After experimenting with production of various algae fuels beginning in 2007, the company now focuses on producing what it calls "green crude" from algae in open raceway ponds. After receiving more than $100 million in federal funds in 2012, Sapphire built the first commercial demonstration algae fuel facility in New Mexico and has continuously produced biofuel since completion of the facility in that year. In 2013, Sapphire began commercial sales of algal biofuel to Tesoro, making it one of the first companies, along with Solazyme, to sell algae fuel on the market.

Diversified Technologies Inc.

Diversified Technologies Inc. has created a patent pending pre-treatment option to reduce costs of oil extraction from algae. This technology, called Pulsed Electric Field (PEF) technology, is a low cost, low energy process that applies high voltage electric pulses to a slurry of algae. The electric pulses enable the algal cell walls to be ruptured easily, increasing the availability of all cell contents (Lipids, proteins and carbohydrates), allowing the separation into specific components downstream. This alternative method to intracellular extraction has shown the capability to be both integrated in-line as well as scalable into high yield assemblies. The Pulse Electric Field subjects the algae to short, intense bursts of electromagnetic radiation in a treatment chamber, electroporating the cell walls. The formation of holes in the cell wall allows the contents within to flow into the surrounding solution for further separation. PEF technology only requires 1-10 microsecond pulses, enabling a high-throughput approach to algal extraction.

Preliminary calculations have shown that utilization of PEF technology would only account for $0.10 per gallon of algae derived biofuel produced. In comparison, conventional drying and solvent-based extractions account for $1.75 per gallon. This inconsistency between costs can be attributed to the fact that algal drying generally accounts for 75% of the extraction process. Although a relatively new technology, PEF has been successfully used in both food decomtamination processes as well as waste water treatments.

Origin Oils Inc.

Origin Oils Inc. has been researching a revolutionary method called the Helix Bioreactor, altering the common closed-loop growth system. This system utilizes low energy lights in a helical pattern, enabling each algal cell to obtain the required amount of light. Sunlight can only penetrate a few inches through algal cells, making light a limiting reagent in open-pond algae farms. Each lighting element in the bioreactor is specially altered to emit specific wavelengths of light, as a full spectrum of light is not beneficial to algae growth. In fact, ultraviolet irradiation is actually detrimental as it inhibits photosynthesis, photoreduction, and the 520 nm light-dark absorbance change of algae.

This bioreactor also addresses another key issue in algal cell growth; introducing CO_2 and nutrients to the algae without disrupting or over-aerating the algae. Origin Oils Inc. combats this issues through the creation of their Quantum Fracturing technology. This process takes the CO_2 and other nutrients, fractures them at extremely high pressures and then deliver the micron sized bubbles to the algae. This allows the nutrients to be delivered at a much lower pressure, maintaining the integrity of the cells.

Proviron

Proviron has been working on a new type of reactor (using flat plates) which reduces the cost of algae cultivation. At AlgaePARC similar research is being conducted using 4 grow systems (1 open pond system and 3 types of closed systems). According to René Wijffels the current systems do not yet allow algae fuel to be produced competitively. However using new (closed) systems, and by scaling up the production it would be possible to reduce costs by 10X, up to a price of 0,4 € per kg of algae.

Genifuels

Genifuel Corporation has licensed the high temperature/pressure fuel extraction process and has been working with the team at the lab since 2008. The company intends to team with some industrial partners to create a pilot plant using this process to make biofuel in industrial quantities. Genifuel process combines hydrothermal liquefaction with catalytic hydrothermal gasification in reactor running at 350 Celsius (662 Fahrenheit) and pressure of 3000 PSI.

Baya® Biofuel

The QMAB started to mobilize its 100 hectares micro algae cultivation farm in Qeshm Island free zone. Development of the farm mainly focuses on 2 phases, production of nutraceutical products and green crude oil to produce biofuel. main product of microalgae culture is crude oil, which can be fractioned into the same kinds of fuels and chemical compounds.

Biodiesel

Bus run by biodiesel.

Biodiesel refers to a vegetable oil - or animal fat-based diesel fuel consisting of long-chain alkyl (methyl, ethyl, or propyl) esters. Biodiesel is typically made by chemically reacting lipids (e.g., vegetable oil, soybean oil, animal fat (tallow)) with an alcohol producing fatty acid esters.

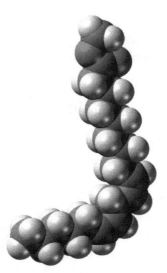

Space-filling model of methyl linoleate, or linoleic acid methyl ester, a common methyl ester produced from soybean or canola oil and methanol

Space-filling model of ethyl stearate, or stearic acid ethyl ester, an ethyl ester produced from soybean or canola oil and ethanol

Biodiesel is meant to be used in standard diesel engines and is thus distinct from the vegetable and waste oils used to fuel *converted* diesel engines. Biodiesel can be used alone, or blended with petrodiesel in any proportions. Biodiesel blends can also be used as heating oil.

The National Biodiesel Board (USA) also has a technical definition of "biodiesel" as a mono-alkyl ester.

Blends

Biodiesel sample

Blends of biodiesel and conventional hydrocarbon-based diesel are products most commonly distributed for use in the retail diesel fuel marketplace. Much of the world uses a system known as the "B" factor to state the amount of biodiesel in any fuel mix:

- 100% biodiesel is referred to as B100

- 20% biodiesel, 80% petrodiesel is labeled B20

- 5% biodiesel, 95% petrodiesel is labeled B5

- 2% biodiesel, 98% petrodiesel is labeled B2

Blends of 20% biodiesel and lower can be used in diesel equipment with no, or only minor modifications, although certain manufacturers do not extend warranty coverage if equipment is damaged by these blends. The B6 to B20 blends are covered by the ASTM D7467 specification. Biodiesel can also be used in its pure form (B100), but may require certain engine modifications to avoid maintenance and performance problems. Blending B100 with petroleum diesel may be accomplished by:

- Mixing in tanks at manufacturing point prior to delivery to tanker truck

- Splash mixing in the tanker truck (adding specific percentages of biodiesel and petroleum diesel)

- In-line mixing, two components arrive at tanker truck simultaneously.

- Metered pump mixing, petroleum diesel and biodiesel meters are set to X total volume, transfer pump pulls from two points and mix is complete on leaving pump.

Applications

Biodiesel can be used in pure form (B100) or may be blended with petroleum diesel at any concentration in most injection pump diesel engines. New extreme high-pressure (29,000 psi) common rail engines have strict factory limits of B5 or B20, depending on manufacturer. Biodiesel has different solvent properties than petrodiesel, and will degrade natural rubber gaskets and hoses in vehicles (mostly vehicles manufactured before 1992), although these tend to wear out naturally and most likely will have already been replaced with FKM, which is nonreactive to biodiesel. Biodiesel has been known to break down deposits of residue in the fuel lines where petrodiesel has been used. As a result, fuel filters may become clogged with particulates if a quick transition to pure biodiesel is made. Therefore, it is recommended to change the fuel filters on engines and heaters shortly after first switching to a biodiesel blend.

Distribution

Since the passage of the Energy Policy Act of 2005, biodiesel use has been increasing in the United States. In the UK, the Renewable Transport Fuel Obligation obliges suppliers to include 5% renewable fuel in all transport fuel sold in the UK by 2010. For road diesel, this effectively means 5% biodiesel (B5).

Vehicular Use and Manufacturer Acceptance

In 2005, Chrysler (then part of DaimlerChrysler) released the Jeep Liberty CRD diesels from the factory into the American market with 5% biodiesel blends, indicating at least partial acceptance of biodiesel as an acceptable diesel fuel additive. In 2007, DaimlerChrysler indicated its intention to increase warranty coverage to 20% biodiesel blends if biofuel quality in the United States can be standardized.

The Volkswagen Group has released a statement indicating that several of its vehicles are compatible with B5 and B100 made from rape seed oil and compatible with the EN 14214 standard. The use of the specified biodiesel type in its cars will not void any warranty.

Mercedes Benz does not allow diesel fuels containing greater than 5% biodiesel (B5) due to concerns about "production shortcomings". Any damages caused by the use of such non-approved fuels will not be covered by the Mercedes-Benz Limited Warranty.

Starting in 2004, the city of Halifax, Nova Scotia decided to update its bus system to allow the fleet of city buses to run entirely on a fish-oil based biodiesel. This caused the city some initial mechanical issues, but after several years of refining, the entire fleet had successfully been converted.

In 2007, McDonalds of UK announced it would start producing biodiesel from the waste oil by-product of its restaurants. This fuel would be used to run its fleet.

The 2014 Chevy Cruze Clean Turbo Diesel, direct from the factory, will be rated for up to B20 (blend of 20% biodiesel / 80% regular diesel) biodiesel compatibility

Railway Usage

Biodiesel locomotive and its external fuel tank at Mount Washington Cog Railway

British train operating company Virgin Trains claimed to have run the UK's first "biodiesel train", which was converted to run on 80% petrodiesel and 20% biodiesel.

The Royal Train on 15 September 2007 completed its first ever journey run on 100% biodiesel fuel supplied by Green Fuels Ltd. His Royal Highness, The Prince of Wales, and Green Fuels managing director, James Hygate, were the first passengers on a train fueled entirely by biodiesel fuel. Since 2007, the Royal Train has operated successfully on B100 (100% biodiesel).

Similarly, a state-owned short-line railroad in eastern Washington ran a test of a 25% biodiesel / 75% petrodiesel blend during the summer of 2008, purchasing fuel from a biodiesel producer sited along the railroad tracks. The train will be powered by biodiesel made in part from canola grown in agricultural regions through which the short line runs.

Also in 2007, Disneyland began running the park trains on B98 (98% biodiesel). The program was discontinued in 2008 due to storage issues, but in January 2009, it was announced that the park would then be running all trains on biodiesel manufactured from its own used cooking oils. This is a change from running the trains on soy-based biodiesel.

In 2007, the historic Mt. Washington Cog Railway added the first biodiesel locomotive to its all-steam locomotive fleet. The fleet has climbed up the western slopes of Mount Washington in New Hampshire since 1868 with a peak vertical climb of 37.4 degrees.

On 8 July 2014, the then Indian Railway Minister D.V. Sadananda Gowda announced in Railway Budget that 5% bio-diesel will be used in Indian Railways' Diesel Engines.

Aircraft Use

A test flight has been performed by a Czech jet aircraft completely powered on biodiesel. Other recent jet flights using biofuel, however, have been using other types of renewable fuels.

On November 7, 2011 United Airlines flew the world's first commercial aviation flight on a microbially derived biofuel using Solajet™, Solazyme's algae-derived renewable jet fuel. The Eco-skies Boeing 737-800 plane was fueled with 40 percent Solajet and 60 percent petroleum-derived jet fuel. The commercial Eco-skies flight 1403 departed from Houston's IAH airport at 10:30 and landed at Chicago's ORD airport at 13:03.

As a Heating Oil

Biodiesel can also be used as a heating fuel in domestic and commercial boilers, a mix of heating oil and biofuel which is standardized and taxed slightly differently from diesel fuel used for transportation. Bioheat® fuel is a proprietary blend of biodiesel and traditional heating oil. Bioheat® is a registered trademark of the National Biodiesel Board [NBB] and the National Oilheat Research Alliance [NORA] in the U.S., and Columbia Fuels in Canada). Heating biodiesel is available in various blends. ASTM 396 recognizes blends of up to 5 percent biodiesel as equivalent to pure petroleum heating oil. Blends of higher levels of up to 20% biofuel are used by many consumers. Research is underway to determine whether such blends affect performance.

Older furnaces may contain rubber parts that would be affected by biodiesel's solvent properties, but can otherwise burn biodiesel without any conversion required. Care must be taken, however, given that varnishes left behind by petrodiesel will be released and can clog pipes- fuel filtering and prompt filter replacement is required. Another approach is to start using biodiesel as a blend, and decreasing the petroleum proportion over time can allow the varnishes to come off more gradually and be less likely to clog. Thanks to its strong solvent properties, however, the furnace is cleaned out and generally becomes more efficient. A technical research paper describes laboratory research and field trials project using pure biodiesel and biodiesel blends as a heating fuel in oil-fired boilers. During the Biodiesel Expo 2006 in the UK, Andrew J. Robertson presented his biodiesel heating oil research from his technical paper and suggested B20 biodiesel could reduce UK household CO_2 emissions by 1.5 million tons per year.

A law passed under Massachusetts Governor Deval Patrick requires all home heating diesel in that state to be 2% biofuel by July 1, 2010, and 5% biofuel by 2013. New York City has passed a similar law.

Cleaning Oil Spills

With 80-90% of oil spill costs invested in shoreline cleanup, there is a search for more efficient and cost-effective methods to extract oil spills from the shorelines. Biodiesel has displayed its capacity to significantly dissolve crude oil, depending on the source of the fatty acids. In a laboratory setting, oiled sediments that simulated polluted shorelines were sprayed with a single coat of biodiesel and exposed to simulated tides. Biodiesel is an effective solvent to oil due to its methyl ester component, which considerably lowers the viscosity of the crude oil. Additionally, it has a higher buoyancy than crude oil, which later aids in its removal. As a result, 80% of oil was removed from cobble and fine sand, 50% in coarse sand, and 30% in gravel. Once the oil is liberated from the shoreline, the oil-biodiesel mixture is manually removed from the water surface with skimmers. Any remaining mixture is easily broken down due to the high biodegradability of biodiesel, and the increased surface area exposure of the mixture.

Biodiesel in Generators

Biodiesel is also used in rental generators

In 2001, UC Riverside installed a 6-megawatt backup power system that is entirely fueled by biodiesel. Backup diesel-fueled generators allow companies to avoid damaging blackouts of critical operations at the expense of high pollution and emission rates. By using B100, these generators were able to essentially eliminate the byproducts that result in smog, ozone, and sulfur emissions. The use of these generators in residential areas around schools, hospitals, and the general public result in substantial reductions in poisonous carbon monoxide and particulate matter.

Historical Background

Transesterification of a vegetable oil was conducted as early as 1853 by Patrick Duffy, four decades before the first diesel engine became functional. Rudolf Diesel's prime model, a single 10 ft (3.0 m) iron cylinder with a flywheel at its base, ran on its own power for the first time in Augsburg, Germany, on 10 August 1893 running on nothing but peanut oil. In remembrance of this event, 10 August has been declared "International Biodiesel Day".

Rudolf Diesel

It is often reported that Diesel designed his engine to run on peanut oil, but this is not the case. Diesel stated in his published papers, "at the Paris Exhibition in 1900 (*Exposition Universelle*) there was shown by the Otto Company a small Diesel engine, which, at the request of the French government ran on arachide (earth-nut or pea-nut) oil, and worked so smoothly that only a few people were aware of it. The engine was constructed for using mineral oil, and was then worked on vegetable oil without any alterations being made. The French Government at the time thought of testing the applicability to power production of the Arachide, or earth-nut, which grows in considerable quantities in their African colonies, and can easily be cultivated there." Diesel himself later conducted related tests and appeared supportive of the idea. In a 1912 speech Diesel said, "the use of vegetable oils for engine fuels may seem insignificant today but such oils may become, in the course of time, as important as petroleum and the coal-tar products of the present time."

Despite the widespread use of petroleum-derived diesel fuels, interest in vegetable oils as fuels for internal combustion engines was reported in several countries during the 1920s and 1930s and later during World War II. Belgium, France, Italy, the United Kingdom, Portugal, Germany, Brazil, Argentina, Japan and China were reported to have tested and used vegetable oils as diesel fuels during this time. Some operational problems were reported due to the high viscosity of vegetable oils compared to petroleum diesel fuel, which results in poor atomization of the fuel in the fuel spray and often leads to deposits and coking of the injectors, combustion chamber and valves. Attempts to overcome these problems included heating of the vegetable oil, blending it with petroleum-derived diesel fuel or ethanol, pyrolysis and cracking of the oils.

On 31 August 1937, G. Chavanne of the University of Brussels (Belgium) was granted a patent for a "Procedure for the transformation of vegetable oils for their uses as fuels" (fr. "*Procédé de Transformation d'Huiles Végétales en Vue de Leur Utilisation comme Carburants*") Belgian Patent 422,877. This patent described the alcoholysis (often referred to as transesterification) of vegetable oils using ethanol (and mentions methanol) in order to separate the fatty acids from the glycerol by replacing the glycerol with short linear alcohols. This appears to be the first account of the production of what is known as "biodiesel" today.

More recently, in 1977, Brazilian scientist Expedito Parente invented and submitted for patent, the first industrial process for the production of biodiesel. This process is classified as biodiesel by international norms, conferring a "standardized identity and quality. No other proposed biofuel has been validated by the motor industry." As of 2010, Parente's company Tecbio is working with Boeing and NASA to certify bioquerosene (bio-kerosene), another product produced and patented by the Brazilian scientist.

Research into the use of transesterified sunflower oil, and refining it to diesel fuel standards, was initiated in South Africa in 1979. By 1983, the process for producing fuel-quality, engine-tested biodiesel was completed and published internationally. An Austrian company, Gaskoks, obtained the technology from the South African Agricultural Engineers; the company erected the first biodiesel pilot plant in November 1987, and the first industrial-scale plant in April 1989 (with a capacity of 30,000 tons of rapeseed per annum).

Throughout the 1990s, plants were opened in many European countries, including the Czech Republic, Germany and Sweden. France launched local production of biodiesel fuel (referred to as *diester*) from rapeseed oil, which is mixed into regular diesel fuel at a level of 5%, and into the diesel fuel used by some captive fleets (e.g. public transportation) at a level of 30%. Renault, Peugeot and other manufacturers have certified truck engines for use with up to that level of partial biodiesel; experiments with 50% biodiesel are underway. During the same period, nations in other parts of the world also saw local production of biodiesel starting up: by 1998, the Austrian Biofuels Institute had identified 21 countries with commercial biodiesel projects. 100% biodiesel is now available at many normal service stations across Europe.

Properties

Biodiesel has promising lubricating properties and cetane ratings compared to low sulfur diesel fuels. Depending on the engine, this might include high pressure injection pumps, pump injectors (also called *unit injectors*) and fuel injectors.

Older diesel Mercedes are popular for running on biodiesel.

The calorific value of biodiesel is about 37.27 MJ/kg. This is 9% lower than regular Number 2 petrodiesel. Variations in biodiesel energy density is more dependent on the feedstock used than the production process. Still, these variations are less than for petrodiesel. It has been claimed biodiesel gives better lubricity and more complete combustion thus increasing the engine energy output and partially compensating for the higher energy density of petrodiesel.

The color of biodiesel ranges from golden to dark brown, depending on the production method. It is slightly miscible with water, has a high boiling point and low vapor pressure. *The flash point of biodiesel (>130 °C, >266 °F) is significantly higher than that of petroleum diesel (64 °C, 147 °F) or gasoline (−45 °C, -52 °F). Biodiesel has a density of ~ 0.88 g/cm³, higher than petrodiesel (~ 0.85 g/cm³).

Biodiesel contains virtually no sulfur, and it is often used as an additive to Ultra-Low Sulfur Diesel (ULSD) fuel to aid with lubrication, as the sulfur compounds in petrodiesel provide much of the lubricity.

Fuel Efficiency

The power output of biodiesel depends on its blend, quality, and load conditions under which the fuel is burnt. The thermal efficiency for example of B100 as compared to B20 will vary due to the differing energy content of the various blends. Thermal efficiency of a fuel is based in part on fuel characteristics such as: viscosity, specific density, and flash point; these characteristics will change as the blends as well as the quality of biodiesel varies. The American Society for Testing and Materials has set standards in order to judge the quality of a given fuel sample.

Regarding brake thermal efficiency one study found that B40 was superior to traditional counterpart at higher compression ratios (this higher brake thermal efficiency was recorded at compression ratios of 21:1). It was noted that, as the compression ratios increased, the efficiency of all fuel types - as well as blends being tested - increased; though it was found that a blend of B40 was the most economical at a compression ratio of 21:1 over all other blends. The study implied that this increase in efficiency was due to fuel density, viscosity, and heating values of the fuels.

Combustion

Fuel systems on the modern diesel engine were not designed to accommodate biodiesel, while many heavy duty engines are able to run with biodiesel blends e.g. B20. Traditional direct injection fuel systems operate at roughly 3,000 psi at the injector tip while the modern common rail fuel system operates upwards of 30,000 PSI at the injector tip. Components are designed to operate at a great temperature range, from below freezing to over 1,000 degrees Fahrenheit. Diesel fuel is expected to burn efficiently and produce as few emissions as possible. As emission standards are being introduced to diesel engines the need to control harmful emissions is being designed into the parameters of diesel engine fuel systems. The traditional inline injection system is more forgiving to poorer quality fuels as opposed to the common rail fuel system. The higher pressures and tighter tolerances of the common rail system allows for greater control over atomization and injection timing. This control of atomization as well as combustion allows for greater efficiency of modern diesel engines as well as greater control over emissions. Components within a diesel fuel system interact with the fuel in a way to ensure efficient operation of the fuel system and so the engine. If an out-of-specification fuel is introduced to a system that has specific parameters of operation, then the integrity of the overall fuel system may be compromised. Some of these parameters such as spray pattern and atomization are directly related to injection timing.

One study found that during atomization biodiesel and its blends produced droplets were greater in diameter than the droplets produced by traditional petrodiesel. The smaller droplets were attributed to the lower viscosity and surface tension of traditional petrol. It was found that droplets at the periphery of the spray pattern were larger in diameter than the droplets at the center this was attributed to the faster pressure drop at the edge of the spray pattern; there was a proportional relationship between the droplet size and the distance from the injector tip. It was found that B100 had the greatest spray penetration, this was attributed to the greater density of B100. Having a greater droplet size can lead to; inefficiencies in the combustion, increased emissions, and decreased horse power. In another study it was found that there is a short injection delay when injecting biodiesel. This injection delay was attributed to the greater viscosity of Biodiesel. It was noted that the higher viscosity and the greater cetane rating of biodiesel over traditional petrodiesel lead to poor atomization, as well as mixture penetration with air during the ignition delay period. Another study noted that this ignition delay may aid in a decrease of NOx emission.

Emissions

Emissions are inherent to the combustion of diesel fuels that are regulated by the U.S. Environmental Protection Agency (E.P.A.). As these emissions are a byproduct of the combustion process, in order to ensure E.P.A. compliance a fuel system must be capable of controlling the combustion of fuels as well as the mitigation of emissions. There are a number of new technologies being phased in to control the production of diesel emissions. The exhaust gas recirculation system, E.G.R., and the diesel particulate filter, D.P.F., are both designed to mitigate the production of harmful emissions.

A study performed by the Chonbuk National University concluded that a B30 biodiesel blend reduced carbon monoxide emissions by approximately 83% and particulate matter emissions by roughly 33%. NOx emissions, however, were found to increase without the application of an E.G.R. system. The study also concluded that, with E.G.R, a B20 biodiesel blend considerably reduced the emissions of the engine. Additionally, analysis by the California Air Resources Board found that biodiesel had the lowest carbon emissions of the fuels tested, those being ultra-low-sulfur diesel, gasoline, corn-based ethanol, compressed natural gas, and five types of biodiesel from varying feedstocks. Their conclusions also showed great variance in carbon emissions of biodiesel based on the feedstock used. Of soy, tallow, canola, corn, and used cooking oil, soy showed the highest carbon emissions, while used cooking oil produced the lowest.

While studying the effect of biodiesel on a D.P.F. it was found that though the presence of sodium and potassium carbonates aided in the catalytic conversion of ash, as the diesel particulates are catalyzed, they may congregate inside the D.P.F. and so interfere with the clearances of the filter. This may cause the filter to clog and interfere with the regeneration process. In a study on the impact of E.G.R. rates with blends of jathropa biodiesel it was shown that there was a decrease in fuel efficiency and torque output due to the use of biodiesel on a diesel engine designed with an E.G.R. system. It was found that CO and CO2 emissions increased with an increase in exhaust gas recirculation but NOx levels decreased. The opacity level of the jathropa blends was in an acceptable range, where traditional diesel was out of acceptable standards. It was shown that a decrease in Nox emissions could be obtained with an E.G.R. system. This study showed an advantage over traditional diesel within a certain operating range of the E.G.R. system. Currently blended biodiesel fuels (B5 and B20) are being used in many heavy-duty vehicles especially transit buses in US cities. Characterization of exhaust emissions showed significant emission reductions compared to regular diesel.

Material Compatibility

- Plastics: High-density polyethylene (HDPE) is compatible but polyvinyl chloride (PVC) is slowly degraded. Polystyrene is dissolved on contact with biodiesel.

- Metals: Biodiesel (like methanol) has an effect on copper-based materials (e.g. brass), and it also affects zinc, tin, lead, and cast iron. Stainless steels (316 and 304) and aluminum are unaffected.

- Rubber: Biodiesel also affects types of natural rubbers found in some older engine components. Studies have also found that fluorinated elastomers (FKM) cured with peroxide and base-metal oxides can be degraded when biodiesel loses its stability caused by oxidation. Commonly used synthetic rubbers FKM- GBL-S and FKM- GF-S found in modern vehicles were found to handle biodiesel in all conditions.

Technical Standards

Biodiesel has a number of standards for its quality including European standard EN 14214, ASTM International D6751, and others.

Low Temperature Gelling

When biodiesel is cooled below a certain point, some of the molecules aggregate and form crystals. The fuel starts to appear cloudy once the crystals become larger than one quarter of the wavelengths of visible light - this is the cloud point (CP). As the fuel is cooled further these crystals become larger. The lowest temperature at which fuel can pass through a 45 micrometre filter is the cold filter plugging point (CFPP). As biodiesel is cooled further it will gel and then solidify. Within Europe, there are differences in the CFPP requirements between countries. This is reflected in the different national standards of those countries. The temperature at which pure (B100) biodiesel starts to gel varies significantly and depends upon the mix of esters and therefore the feedstock oil used to produce the biodiesel. For example, biodiesel produced from low erucic acid varieties of canola seed (RME) starts to gel at approximately −10 °C (14 °F). Biodiesel produced from beef tallow and palm oil tends to gel at around 16 °C (61 °F) and 13 °C (55 °F) respectively. There are a number of commercially available additives that will significantly lower the pour point and cold filter plugging point of pure biodiesel. Winter operation is also possible by blending biodiesel with other fuel oils including #2 low sulfur diesel fuel and #1 diesel / kerosene.

Another approach to facilitate the use of biodiesel in cold conditions is by employing a second fuel tank for biodiesel in addition to the standard diesel fuel tank. The second fuel tank can be insulated and a heating coil using engine coolant is run through the tank. The fuel tanks can be switched over when the fuel is sufficiently warm. A similar method can be used to operate diesel vehicles using straight vegetable oil.

Contamination by Water

Biodiesel may contain small but problematic quantities of water. Although it is only slightly miscible with water it is hygroscopic. One of the reasons biodiesel can absorb water is the persistence of mono and diglycerides left over from an incomplete reaction. These molecules can act as an emulsifier, allowing water to mix with the biodiesel. In addition, there may be water that is residual to processing or resulting from storage tank condensation. The presence of water is a problem because:

- Water reduces the heat of fuel combustion, causing smoke, harder starting, and reduced power.

- Water causes corrosion of fuel system components (pumps, fuel lines, etc.)

- Microbes in water cause the paper-element filters in the system to rot and fail, causing failure of the fuel pump due to ingestion of large particles.

- Water freezes to form ice crystals that provide sites for nucleation, accelerating gelling of the fuel.

- Water causes pitting in pistons.

Previously, the amount of water contaminating biodiesel has been difficult to measure by taking samples, since water and oil separate. However, it is now possible to measure the water content using water-in-oil sensors.

Water contamination is also a potential problem when using certain chemical catalysts involved in the production process, substantially reducing catalytic efficiency of base (high pH) catalysts such as potassium hydroxide. However, the super-critical methanol production methodology, whereby the transesterification process of oil feedstock and methanol is effectuated under high temperature and pressure, has been shown to be largely unaffected by the presence of water contamination during the production phase.

Availability and Prices

In some countries biodiesel is less expensive than conventional diesel

Global biodiesel production reached 3.8 million tons in 2005. Approximately 85% of biodiesel production came from the European Union.

In 2007, in the United States, average retail (at the pump) prices, including federal and state fuel taxes, of B2/B5 were lower than petroleum diesel by about 12 cents, and B20 blends were the same as petrodiesel. However, as part of a dramatic shift in diesel pricing, by July 2009, the US DOE was reporting average costs of B20 15 cents per gallon higher than petroleum diesel ($2.69/gal vs. $2.54/gal). B99 and B100 generally cost more than petrodiesel except where local governments provide a tax incentive or subsidy.

Production

Biodiesel is commonly produced by the transesterification of the vegetable oil or animal fat feedstock. There are several methods for carrying out this transesterification reaction including the common batch process, supercritical processes, ultrasonic methods, and even microwave methods.

Chemically, transesterified biodiesel comprises a mix of mono-alkyl esters of long chain fatty acids. The most common form uses methanol (converted to sodium methoxide) to produce methyl esters (commonly referred to as Fatty Acid Methyl Ester - FAME) as it is the cheapest alcohol available, though ethanol can be used to produce an ethyl ester (commonly referred to as Fatty Acid Ethyl Ester - FAEE) biodiesel and higher alcohols such as isopropanol and butanol have also been used. Using alcohols of higher molecular weights improves the cold flow properties of the resulting ester, at the cost of a less efficient transesterification reaction. A lipid transesterification production process is used to convert the base oil to the desired esters. Any free fatty acids (FFAs) in the base oil are either converted to soap and removed from the process, or they are esterified (yielding more biodiesel) using an acidic catalyst. After this processing, unlike straight vegetable oil, biodiesel has combustion properties very similar to those of petroleum diesel, and can replace it in most current uses.

The methanol used in most biodiesel production processes is made using fossil fuel inputs. However, there are sources of renewable methanol made using carbon dioxide or biomass as feedstock, making their production processes free of fossil fuels.

A by-product of the transesterification process is the production of glycerol. For every 1 tonne of biodiesel that is manufactured, 100 kg of glycerol are produced. Originally, there was a valuable market for the glycerol, which assisted the economics of the process as a whole. However, with the increase in global biodiesel production, the market price for this crude glycerol (containing 20% water and catalyst residues) has crashed. Research is being conducted globally to use this glycerol as a chemical building block. One initiative in the UK is The Glycerol Challenge.

Usually this crude glycerol has to be purified, typically by performing vacuum distillation. This is rather energy intensive. The refined glycerol (98%+ purity) can then be utilised directly, or converted into other products. The following announcements were made in 2007: A joint venture of Ashland Inc. and Cargill announced plans to make propylene glycol in Europe from glycerol and Dow Chemical announced similar plans for North America. Dow also plans to build a plant in China to make epichlorhydrin from glycerol. Epichlorhydrin is a raw material for epoxy resins.

Production Levels

In 2007, biodiesel production capacity was growing rapidly, with an average annual growth rate from 2002-06 of over 40%. For the year 2006, the latest for which actual production figures could be obtained, total world biodiesel production was about 5-6 million tonnes, with 4.9 million tonnes processed in Europe (of which 2.7 million tonnes was from Germany) and most of the rest from the USA. In 2008 production in Europe alone had risen to 7.8 million tonnes. In July 2009, a duty was added to American imported biodiesel in the European Union in order to balance the competition from European, especially German producers. The capacity for 2008 in Europe totalled 16 million tonnes. This compares with a total demand for diesel in the US and Europe of approximately 490 million tonnes (147 billion gallons). Total world production of vegetable oil for all purposes in 2005/06 was about 110 million tonnes, with about 34 million tonnes each of palm oil and soybean oil.

US biodiesel production in 2011 brought the industry to a new milestone. Under the EPA Renewable Fuel Standard, targets have been implemented for the biodiesel production plants in order to monitor and document production levels in comparison to total demand. According to the year-end data released by the EPA, biodiesel production in 2011 reached more than 1 billion gallons. This production number far exceeded the 800 million gallon target set by the EPA. The projected production for 2020 is nearly 12 billion gallons.

Biodiesel Feedstocks

A variety of oils can be used to produce biodiesel. These include:

- Virgin oil feedstock – rapeseed and soybean oils are most commonly used, soybean oil accounting for about half of U.S. production. It also can be obtained from Pongamia, field pennycress and jatropha and other crops such as mustard, jojoba, flax, sunflower, palm oil, coconut and hemp;

- Waste vegetable oil (WVO);

- Animal fats including tallow, lard, yellow grease, chicken fat, and the by-products of the production of Omega-3 fatty acids from fish oil.

- Algae, which can be grown using waste materials such as sewage and without displacing land currently used for food production.

- Oil from halophytes such as *Salicornia bigelovii*, which can be grown using saltwater in coastal areas where conventional crops cannot be grown, with yields equal to the yields of soybeans and other oilseeds grown using freshwater irrigation

- Sewage Sludge - The sewage-to-biofuel field is attracting interest from major companies like Waste Management and startups like InfoSpi, which are betting that renewable sewage biodiesel can become competitive with petroleum diesel on price.

Many advocates suggest that waste vegetable oil is the best source of oil to produce biodiesel, but since the available supply is drastically less than the amount of petroleum-based fuel that is burned for transportation and home heating in the world, this local solution could not scale to the current rate of consumption.

Animal fats are a by-product of meat production and cooking. Although it would not be efficient to raise animals (or catch fish) simply for their fat, use of the by-product adds value to the livestock industry (hogs, cattle, poultry). Today, multi-feedstock biodiesel facilities are producing high quality animal-fat based biodiesel. Currently, a 5-million dollar plant is being built in the USA, with the intent of producing 11.4 million litres (3 million gallons) biodiesel from some of the estimated 1 billion kg (2.2 billion pounds) of chicken fat produced annually at the local Tyson poultry plant. Similarly, some small-scale biodiesel factories use waste fish oil as feedstock. An EU-funded project (ENERFISH) suggests that at a Vietnamese plant to produce biodiesel from catfish (basa, also known as pangasius), an output of 13 tons/day of biodiesel can be produced from 81 tons of fish waste (in turn resulting from 130 tons of fish). This project utilises the biodiesel to fuel a CHP unit in the fish processing plant, mainly to power the fish freezing plant.

Quantity of Feedstocks Required

Current worldwide production of vegetable oil and animal fat is not sufficient to replace liquid fossil fuel use. Furthermore, some object to the vast amount of farming and the resulting fertilization, pesticide use, and land use conversion that would be needed to produce the additional vegetable oil. The estimated transportation diesel fuel and home heating oil used in the United States is about 160 million tons (350 billion pounds) according to the Energy Information Administration, US Department of Energy. In the United States, estimated production of vegetable oil for all uses is about 11 million tons (24 billion pounds) and estimated production of animal fat is 5.3 million tonnes (12 billion pounds).

If the entire arable land area of the USA (470 million acres, or 1.9 million square kilometers) were devoted to biodiesel production from soy, this would just about provide the 160 million tonnes required (assuming an optimistic 98 US gal/acre of biodiesel). This land area could in principle be reduced significantly using algae, if the obstacles can be overcome. The US DOE estimates that if algae fuel replaced all the petroleum fuel in the United States, it would require 15,000 square miles (39,000 square kilometers), which is a few thousand square miles larger than Maryland, or 30% greater than the area of Belgium, assuming a yield of 140 tonnes/hectare (15,000 US gal/ acre). Given a more realistic yield of 36 tonnes/hectare (3834 US gal/acre) the area required is about 152,000 square kilometers, or roughly equal to that of the state of Georgia or of England and Wales. The advantages of algae are that it can be grown on non-arable land such as deserts or in marine environments, and the potential oil yields are much higher than from plants.

Yield

Feedstock yield efficiency per unit area affects the feasibility of ramping up production to the huge industrial levels required to power a significant percentage of vehicles.

Some typical yields		
Crop	Yield	
	L/ha	US gal/acre
Palm oil	4752	508
Coconut	2151	230
Cyperus esculentus	1628	174
Rapeseed	954	102
Soy (Indiana)	554-922	59.2-98.6
Chinese tallow	907	97
Peanut	842	90
Sunflower	767	82
Hemp	242	26

1. "Biofuels: some numbers". Grist.org. Retrieved 2010-03-15.
2. Makareviciene et al., "Opportunities for the use of chufa sedge in biodiesel production", Industrial Crops and Products, 50 (2013) p. 635, table 2.
3. Klass, Donald, "Biomass for Renewable Energy, Fuels, and Chemicals", page 341. Academic Press, 1998.
4. Kitani, Osamu, "Volume V: Energy and Biomass Engineering, CIGR Handbook of Agricultural Engineering", Amer Society of Agricultural, 1999.

Algae fuel yields have not yet been accurately determined, but DOE is reported as saying that algae yield 30 times more energy per acre than land crops such as soybeans. Yields of 36 tonnes/hectare are considered practical by Ami Ben-Amotz of the Institute of Oceanography in Haifa, who has been farming Algae commercially for over 20 years.

Jatropha has been cited as a high-yield source of biodiesel but yields are highly dependent on climatic and soil conditions. The estimates at the low end put the yield at about 200 US gal/ acre (1.5-2 tonnes per hectare) per crop; in more favorable climates two or more crops per year have been achieved. It is grown in the Philippines, Mali and India, is drought-resistant, and can share space with other cash crops such as coffee, sugar, fruits and vegetables. It is well-suited to semi-arid lands and can contribute to slow down desertification, according to its advocates.

Efficiency and Economic Arguments

Pure biodiesel (B-100) made from soybeans

According to a study by Drs. Van Dyne and Raymer for the Tennessee Valley Authority, the average US farm consumes fuel at the rate of 82 litres per hectare (8.75 US gal/acre) of land to produce one crop. However, average crops of rapeseed produce oil at an average rate of 1,029 L/ha (110 US gal/acre), and high-yield rapeseed fields produce about 1,356 L/ha (145 US gal/acre). The ratio of input to output in these cases is roughly 1:12.5 and 1:16.5. Photosynthesis is known to have an efficiency rate of about 3-6% of total solar radiation and if the entire mass of a crop is utilized for energy production, the overall efficiency of this chain is currently about 1% While this may compare unfavorably to solar cells combined with an electric drive train, biodiesel is less costly to deploy (solar cells cost approximately US$250 per square meter) and transport (electric vehicles require batteries which currently have a much lower energy density than liquid fuels). A 2005 study found that biodiesel production using soybeans required 27% more fossil energy than the biodiesel produced and 118% more energy using sunflowers.

However, these statistics by themselves are not enough to show whether such a change makes economic sense. Additional factors must be taken into account, such as: the fuel equivalent of the energy required for processing, the yield of fuel from raw oil, the return on cultivating food, the effect biodiesel will have on food prices and the relative cost of biodiesel versus petrodiesel, water pollution from farm run-off, soil depletion, and the externalized costs of political and military interference in oil-producing countries intended to control the price of petrodiesel.

The debate over the energy balance of biodiesel is ongoing. Transitioning fully to biofuels could require immense tracts of land if traditional food crops are used (although non food crops can be utilized). The problem would be especially severe for nations with large economies, since energy consumption scales with economic output.

If using only traditional food plants, most such nations do not have sufficient arable land to produce biofuel for the nation's vehicles. Nations with smaller economies (hence less energy consumption) and more arable land may be in better situations, although many regions cannot afford to divert land away from food production.

For third world countries, biodiesel sources that use marginal land could make more sense; e.g., pongam oiltree nuts grown along roads or jatropha grown along rail lines.

In tropical regions, such as Malaysia and Indonesia, plants that produce palm oil are being planted at a rapid pace to supply growing biodiesel demand in Europe and other markets. Scientists have shown that the removal of rainforest for palm plantations is not ecologically sound since the expansion of oil palm plantations poses a threat to natural rainforest and biodiversity.

It has been estimated in Germany that palm oil biodiesel has less than one third of the production costs of rapeseed biodiesel. The direct source of the energy content of biodiesel is solar energy captured by plants during photosynthesis. Regarding the positive energy balance of biodiesel:

When straw was left in the field, biodiesel production was strongly energy positive, yielding 1 GJ biodiesel for every 0.561 GJ of energy input (a yield/cost ratio of 1.78).

> When straw was burned as fuel and oilseed rapemeal was used as a fertilizer, the yield/cost ratio for biodiesel production was even better (3.71). In other words, for every unit of energy input to produce biodiesel, the output was 3.71 units (the difference of 2.71 units would be from solar energy).

Economic Impact

Multiple economic studies have been performed regarding the economic impact of biodiesel production. One study, commissioned by the National Biodiesel Board, reported the 2011 production of biodiesel supported 39,027 jobs and more than $2.1 billion in household income. The growth in biodiesel also helps significantly increase GDP. In 2011, biodiesel created more than $3 billion in GDP. Judging by the continued growth in the Renewable Fuel Standard and the extension of the biodiesel tax incentive, the number of jobs can increase to 50,725, $2.7 billion in income, and reaching $5 billion in GDP by 2012 and 2013.

Energy Security

One of the main drivers for adoption of biodiesel is energy security. This means that a nation's dependence on oil is reduced, and substituted with use of locally available sources, such as coal, gas, or renewable sources. Thus a country can benefit from adoption of biofuels, without a reduction in greenhouse gas emissions. While the total energy balance is debated, it is clear that the dependence on oil is reduced. One example is the energy used to manufacture fertilizers, which could come from a variety of sources other than petroleum. The US National Renewable Energy Laboratory (NREL) states that energy security is the number one driving force behind the US biofuels programme, and a White House "Energy Security for the 21st Century" paper makes it clear that energy security is a major reason for promoting biodiesel. The former EU commission president, Jose Manuel Barroso, speaking at a recent EU biofuels conference, stressed that properly managed biofuels have the potential to reinforce the EU's security of supply through diversification of energy sources.

Global Biofuel Policies

Many countries around the world are involved in the growing use and production of biofuels, such as biodiesel, as an alternative energy source to fossil fuels and oil. To foster the biofuel industry, governments have implemented legislations and laws as incentives to reduce oil dependency and to increase the use of renewable energies. Many countries have their own independent policies regarding the taxation and rebate of biodiesel use, import, and production.

Canada

It was required by the Canadian Environmental Protection Act Bill C-33 that by the year 2010, gasoline contained 5% renewable content and that by 2013, diesel and heating oil contained 2% renewable content. The EcoENERGY for Biofuels Program subsidized the production of biodiesel, among other biofuels, via an incentive rate of CAN$0.20 per liter from 2008 to 2010. A decrease of $0.04 will be applied every year following, until the incentive rate reaches $0.06 in 2016. Individual provinces also have specific legislative measures in regards to biofuel use and production.

United States

The Volumetric Ethanol Excise Tax Credit (VEETC) was the main source of financial support for biofuels, but was scheduled to expire in 2010. Through this act, biodiesel production guaranteed a tax credit of US$1 per gallon produced from virgin oils, and $0.50 per gallon made from recycled oils. Currently soybean oil is being used to produce soybean biodiesel for many commercial purposes such as blending fuel for transportation sectors.

European Union

The European Union is the greatest producer of biodiesel, with France and Germany being the top producers. To increase the use of biodiesel, there exist policies requiring the blending of biodiesel into fuels, including penalties if those rates are not reached. In France, the goal was to reach 10% integration but plans for that stopped in 2010. As an incentive for the European Union countries to continue the production of the biofuel, there are tax rebates for specific quotas of biofuel produced. In Germany, the minimum percentage of biodiesel in transport diesel is set at 7% so called "B7".

Environmental Effects

The surge of interest in biodiesels has highlighted a number of environmental effects associated with its use. These potentially include reductions in greenhouse gas emissions, deforestation, pollution and the rate of biodegradation.

According to the EPA's Renewable Fuel Standards Program Regulatory Impact Analysis, released in February 2010, biodiesel from soy oil results, on average, in a 57% reduction in greenhouse gases compared to petroleum diesel, and biodiesel produced from waste grease results in an 86% reduction.

However, environmental organizations, for example, Rainforest Rescue and Greenpeace, criticize the cultivation of plants used for biodiesel production, e.g., oil palms, soybeans and sugar cane.

They say the deforestation of rainforests exacerbates climate change and that sensitive ecosystems are destroyed to clear land for oil palm, soybean and sugar cane plantations. Moreover, that biofuels contribute to world hunger, seeing as arable land is no longer used for growing foods. The Environmental Protection Agency(EPA) published data in January 2012, showing that biofuels made from palm oil won't count towards the nation's renewable fuels mandate as they are not climate-friendly. Environmentalists welcome the conclusion because the growth of oil palm plantations has driven tropical deforestation, for example, in Indonesia and Malaysia.

Food, Land and Water vs. Fuel

In some poor countries the rising price of vegetable oil is causing problems. Some propose that fuel only be made from non-edible vegetable oils such as camelina, jatropha or seashore mallow which can thrive on marginal agricultural land where many trees and crops will not grow, or would produce only low yields.

Others argue that the problem is more fundamental. Farmers may switch from producing food crops to producing biofuel crops to make more money, even if the new crops are not edible. The law of supply and demand predicts that if fewer farmers are producing food the price of food will rise. It may take some time, as farmers can take some time to change which things they are growing, but increasing demand for first generation biofuels is likely to result in price increases for many kinds of food. Some have pointed out that there are poor farmers and poor countries who are making more money because of the higher price of vegetable oil.

Biodiesel from sea algae would not necessarily displace terrestrial land currently used for food production and new algaculture jobs could be created.

Current Research

There is ongoing research into finding more suitable crops and improving oil yield. Other sources are possible including human fecal matter, with Ghana building its first "fecal sludge-fed biodiesel plant." Using the current yields, vast amounts of land and fresh water would be needed to produce enough oil to completely replace fossil fuel usage. It would require twice the land area of the US to be devoted to soybean production, or two-thirds to be devoted to rapeseed production, to meet current US heating and transportation needs.

Specially bred mustard varieties can produce reasonably high oil yields and are very useful in crop rotation with cereals, and have the added benefit that the meal leftover after the oil has been pressed out can act as an effective and biodegradable pesticide.

The NFESC, with Santa Barbara-based Biodiesel Industries is working to develop biodiesel technologies for the US navy and military, one of the largest diesel fuel users in the world.

A group of Spanish developers working for a company called Ecofasa announced a new biofuel made from trash. The fuel is created from general urban waste which is treated by bacteria to produce fatty acids, which can be used to make biodiesel.

Another approach that does not require the use of chemical for the production involves the use of genetically modified microbes.

Algal Biodiesel

From 1978 to 1996, the U.S. NREL experimented with using algae as a biodiesel source in the "Aquatic Species Program". A self-published article by Michael Briggs, at the UNH Biodiesel Group, offers estimates for the realistic replacement of all vehicular fuel with biodiesel by utilizing algae that have a natural oil content greater than 50%, which Briggs suggests can be grown on algae ponds at wastewater treatment plants. This oil-rich algae can then be extracted from the system and processed into biodiesel, with the dried remainder further reprocessed to create ethanol.

The production of algae to harvest oil for biodiesel has not yet been undertaken on a commercial scale, but feasibility studies have been conducted to arrive at the above yield estimate. In addition to its projected high yield, algaculture — unlike crop-based biofuels — does not entail a decrease in food production, since it requires neither farmland nor fresh water. Many companies are pursuing algae bio-reactors for various purposes, including scaling up biodiesel production to commercial levels.

Prof. Rodrigo E. Teixeira from the University of Alabama in Huntsville demonstrated the extraction of biodiesel lipids from wet algae using a simple and economical reaction in ionic liquids.

Pongamia

Millettia pinnata, also known as the Pongam Oiltree or Pongamia, is a leguminous, oilseed-bearing tree that has been identified as a candidate for non-edible vegetable oil production.

Pongamia plantations for biodiesel production have a two-fold environmental benefit. The trees both store carbon and produce fuel oil. Pongamia grows on marginal land not fit for food crops and does not require nitrate fertilizers. The oil producing tree has the highest yield of oil producing plant (approximately 40% by weight of the seed is oil) while growing in malnourished soils with high levels of salt. It is becoming a main focus in a number of biodiesel research organizations. The main advantages of Pongamia are a higher recovery and quality of oil than other crops and no direct competition with food crops. However, growth on marginal land can lead to lower oil yields which could cause competition with food crops for better soil.

Jatropha

Jatropha Biodiesel from DRDO, India.

Several groups in various sectors are conducting research on Jatropha curcas, a poisonous shrub-like tree that produces seeds considered by many to be a viable source of biodiesel feedstock oil. Much of this research focuses on improving the overall per acre oil yield of Jatropha through advancements in genetics, soil science, and horticultural practices.

SG Biofuels, a San Diego-based Jatropha developer, has used molecular breeding and biotechnology to produce elite hybrid seeds of Jatropha that show significant yield improvements over first generation varieties. SG Biofuels also claims that additional benefits have arisen from such strains, including improved flowering synchronicity, higher resistance to pests and disease, and increased cold weather tolerance.

Plant Research International, a department of the Wageningen University and Research Centre in the Netherlands, maintains an ongoing Jatropha Evaluation Project (JEP) that examines the feasibility of large scale Jatropha cultivation through field and laboratory experiments.

The Center for Sustainable Energy Farming (CfSEF) is a Los Angeles-based non-profit research organization dedicated to Jatropha research in the areas of plant science, agronomy, and horticulture. Successful exploration of these disciplines is projected to increase Jatropha farm production yields by 200-300% in the next ten years.

Fungi

A group at the Russian Academy of Sciences in Moscow published a paper in September 2008, stating that they had isolated large amounts of lipids from single-celled fungi and turned it into biodiesel in an economically efficient manner. More research on this fungal species; *Cunninghamella japonica*, and others, is likely to appear in the near future.

The recent discovery of a variant of the fungus *Gliocladium roseum* points toward the production of so-called myco-diesel from cellulose. This organism was recently discovered in the rainforests of northern Patagonia and has the unique capability of converting cellulose into medium length hydrocarbons typically found in diesel fuel.

Biodiesel from Used Coffee Grounds

Researchers at the University of Nevada, Reno, have successfully produced biodiesel from oil derived from used coffee grounds. Their analysis of the used grounds showed a 10% to 15% oil content (by weight). Once the oil was extracted, it underwent conventional processing into biodiesel. It is estimated that finished biodiesel could be produced for about one US dollar per gallon. Further, it was reported that "the technique is not difficult" and that "there is so much coffee around that several hundred million gallons of biodiesel could potentially be made annually." However, even if all the coffee grounds in the world were used to make fuel, the amount produced would be less than 1 percent of the diesel used in the United States annually. "It won't solve the world's energy problem," Dr. Misra said of his work.

Exotic Sources

Recently, alligator fat was identified as a source to produce biodiesel. Every year, about 15 million pounds of alligator fat are disposed of in landfills as a waste byproduct of the alligator meat and skin industry. Studies have shown that biodiesel produced from alligator fat is similar in composition to biodiesel created from soybeans, and is cheaper to refine since it is primarily a waste product.

Biodiesel to Hydrogen-cell Power

A microreactor has been developed to convert biodiesel into hydrogen steam to power fuel cells.

Steam reforming, also known as fossil fuel reforming is a process which produces hydrogen gas from hydrocarbon fuels, most notably biodiesel due to its efficiency. A **microreactor**, or reformer, is the processing device in which water vapour reacts with the liquid fuel under high temperature and pressure. Under temperatures ranging from 700 – 1100 °C, a nickel-based catalyst enables the production of carbon monoxide and hydrogen:

$$Hydrocarbon + H2O \rightleftharpoons CO + 3\ H2\ (Highly\ endothermic)$$

Furthermore, a higher yield of hydrogen gas can be harnessed by further oxidizing carbon monoxide to produce more hydrogen and carbon dioxide:

$$CO + H2O \rightarrow CO2 + H2\ (Mildly\ exothermic)$$

Hydrogen Fuel Cells Background Information

Fuel cells operate similar to a battery in that electricity is harnessed from chemical reactions. The difference in fuel cells when compared to batteries is their ability to be powered by the constant flow of hydrogen found in the atmosphere. Furthermore, they produce only water as a by-product, and are virtually silent. The downside of hydrogen powered fuel cells is the high cost and dangers of storing highly combustible hydrogen under pressure.

One way new processors can overcome the dangers of transporting hydrogen is to produce it as necessary. The microreactors can be joined to create a system that heats the hydrocarbon under high pressure to generate hydrogen gas and carbon dioxide, a process called steam reforming. This produces up to 160 gallons of hydrogen/minute and gives the potential of powering hydrogen refueling stations, or even an on-board hydrogen fuel source for hydrogen cell vehicles. Implementation into cars would allow energy-rich fuels, such as biodiesel, to be transferred to kinetic energy while avoiding combustion and pollutant byproducts. The hand-sized square piece of metal contains microscopic channels with catalytic sites, which continuously convert biodiesel, and even its glycerol byproduct, to hydrogen.

Concerns

Engine Wear

Lubricity of fuel plays an important role in wear that occurs in an engine. An engine relies on its fuel to provide lubricity for the metal components that are constantly in contact with each other. Biodiesel is a much better lubricant compared with petroleum diesel due to the presence of esters. Tests have shown that the addition of a small amount of biodiesel to diesel can significantly increase the lubricity of the fuel in short term. However, over a longer period of time (2–4 years), studies show that biodiesel loses its lubricity. This could be because of enhanced corrosion over time due to oxidation of the unsaturated molecules or increased water content in biodiesel from moisture absorption.

Fuel Viscosity

One of the main concerns regarding biodiesel is its viscosity. The viscosity of diesel is 2.5–3.2 cSt at 40 °C and the viscosity of biodiesel made from soybean oil is between 4.2 and 4.6 cSt The viscosity of diesel must be high enough to provide sufficient lubrication for the engine parts but low enough to flow at operational temperature. High viscosity can plug the fuel filter and injection system in engines. Vegetable oil is composed of lipids with long chains of hydrocarbons, to reduce its viscosity the lipids are broken down into smaller molecules of esters. This is done by converting vegetable oil and animal fats into alkyl esters using transesterification to reduce their viscosity Nevertheless, biodiesel viscosity remains higher than that of diesel, and the engine may not be able to use the fuel at low temperatures due to the slow flow through the fuel filter.

Engine Performance

Biodiesel has higher brake-specific fuel consumption compared to diesel, which means more biodiesel fuel consumption is required for the same torque. However, B20 biodiesel blend has been found to provide maximum increase in thermal efficiency, lowest brake-specific energy consumption, and lower harmful emissions. The engine performance depends on the properties of the fuel, as well as on combustion, injector pressure and many other factors. Since there are various blends of biodiesel, that may account for the contradicting reports in regards engine performance.

Biogas

Pipes carrying biogas (foreground), natural gas and condensate

Biogas typically refers to a mixture of different gases produced by the breakdown of organic matter in the absence of oxygen. Biogas can be produced from raw materials such as agricultural waste, manure, municipal waste, plant material, sewage, green waste or food waste. Biogas is a renewable energy source and in many cases exerts a very small carbon footprint.

Biogas can be produced by anaerobic digestion with anaerobic organisms, which digest material inside a closed system, or fermentation of biodegradable materials.

Biogas is primarily methane (CH_4) and carbon dioxide (CO_2) and may have small amounts of hydrogen sulfide (H_2S), moisture and siloxanes. The gases methane, hydrogen, and carbon monoxide (CO) can be combusted or oxidized with oxygen. This energy release allows biogas to be used as a fuel; it can be used for any heating purpose, such as cooking. It can also be used in a gas engine to convert the energy in the gas into electricity and heat.

Biogas can be compressed, the same way natural gas is compressed to CNG, and used to power motor vehicles. In the UK, for example, biogas is estimated to have the potential to replace around 17% of vehicle fuel. It qualifies for renewable energy subsidies in some parts of the world. Biogas can be cleaned and upgraded to natural gas standards, when it becomes bio-methane. Biogas is considered to be a renewable resource because its production-and-use cycle is continuous, and it generates no net carbon dioxide. Organic material grows, is converted and used and then regrows in a continually repeating cycle. From a carbon perspective, as much carbon dioxide is absorbed from the atmosphere in the growth of the primary bio-resource as is released when the material is ultimately converted to energy.

Production

Biogas production in rural Germany

Biogas is produced as landfill gas (LFG), which is produced by the breakdown of Biodegradable waste inside a landfill due to chemical reactions and microbes, or as digested gas, produced inside an anaerobic digester. A *biogas plant* is the name often given to an anaerobic digester that treats farm wastes or energy crops. It can be produced using anaerobic digesters (air-tight tanks with different configurations). These plants can be fed with energy crops such as maize silage or biodegradable wastes including sewage sludge and food waste. During the process, the microorganisms transform biomass waste into biogas (mainly methane and carbon dioxide) and digestate. The biogas is a renewable energy that can be used for heating, electricity, and many other operations that use a reciprocating internal combustion engine, such as GE Jenbacher or Caterpillar gas engines. Other internal combustion engines such as gas turbines are suitable for the conversion of biogas into both electricity and heat. The digestate is the remaining inorganic matter that was not transformed into biogas. It can be used as an agricultural fertiliser.

There are two key processes: mesophilic and thermophilic digestion which is dependent on temperature. In experimental work at University of Alaska Fairbanks, a 1000-litre digester using psychrophiles harvested from "mud from a frozen lake in Alaska" has produced 200–300 liters of methane per day, about 20%–30% of the output from digesters in warmer climates.

Dangers

The dangers of biogas are mostly similar to those of natural gas, but with an additional risk from the toxicity of its hydrogen sulfide fraction. Biogas can be explosive when mixed in the ratio of one part biogas to 8-20 parts air. Special safety precautions have to be taken for entering an empty biogas digester for maintenance work.

It is important that a biogas system never has negative pressure as this could cause an explosion. Negative gas pressure can occur if too much gas is removed or leaked; Because of this biogas should not be used at pressures below one column inch of water, measured by a pressure gauge.

Frequent smell checks must be performed on a biogas system. If biogas is smelled anywhere windows and doors should be opened immediately. If there is a fire the gas should be shut off at the gate valve of the biogas system.

Landfill Gas

Landfill gas is produced by wet organic waste decomposing under anaerobic conditions in a biogas.

The waste is covered and mechanically compressed by the weight of the material that is deposited above. This material prevents oxygen exposure thus allowing anaerobic microbes to thrive. This gas builds up and is slowly released into the atmosphere if the site has not been engineered to capture the gas. Landfill gas released in an uncontrolled way can be hazardous since it can become explosive when it escapes from the landfill and mixes with oxygen. The lower explosive limit is 5% methane and the upper is 15% methane.

The methane in biogas is 20 times more potent a greenhouse gas than carbon dioxide. Therefore, uncontained landfill gas, which escapes into the atmosphere may significantly contribute to the effects of global warming. In addition, volatile organic compounds (VOCs) in landfill gas contribute to the formation of photochemical smog.

Technical

Biochemical oxygen demand (BOD) is a measure of the amount of oxygen required by aerobic micro-organisms to decompose the organic matter in a sample of water. Knowing the energy density of the material being used in the biodigester as well as the BOD for the liquid discharge allows for the calculation of the daily energy output from a biodigester.

Another term related to biodigesters is effluent dirtiness, which tells how much organic material there is per unit of biogas source. Typical units for this measure are in mg BOD/litre. As an example, effluent dirtiness can range between 800–1200 mg BOD/litre in Panama.

From 1 kg of decommissioned kitchen bio-waste, 0.45 m³ of biogas can be obtained. The price for collecting biological waste from households is approximately €70 per ton.

Composition

Typical composition of biogas		
Compound	Formula	%
Methane	CH_4	50–75
Carbon dioxide	CO_2	25–50

Nitrogen	N_2	0–10
Hydrogen	H_2	0–1
Hydrogen sulfide	H_2S	0–3
Oxygen	O_2	0–0.5

The composition of biogas varies depending upon the origin of the anaerobic digestion process. Landfill gas typically has methane concentrations around 50%. Advanced waste treatment technologies can produce biogas with 55%–75% methane, which for reactors with free liquids can be increased to 80%-90% methane using in-situ gas purification techniques. As produced, biogas contains water vapor. The fractional volume of water vapor is a function of biogas temperature; correction of measured gas volume for water vapor content and thermal expansion is easily done via simple mathematics which yields the standardized volume of dry biogas.

In some cases, biogas contains siloxanes. They are formed from the anaerobic decomposition of materials commonly found in soaps and detergents. During combustion of biogas containing siloxanes, silicon is released and can combine with free oxygen or other elements in the combustion gas. Deposits are formed containing mostly silica ($SiO2$) or silicates (Si_xO_y) and can contain calcium, sulfur, zinc, phosphorus. Such *white mineral* deposits accumulate to a surface thickness of several millimeters and must be removed by chemical or mechanical means.

Practical and cost-effective technologies to remove siloxanes and other biogas contaminants are available.

For 1000 kg (wet weight) of input to a typical biodigester, total solids may be 30% of the wet weight while volatile suspended solids may be 90% of the total solids. Protein would be 20% of the volatile solids, carbohydrates would be 70% of the volatile solids, and finally fats would be 10% of the volatile solids.

Benefits of Manure Derived Biogas

High levels of methane are produced when manure is stored under anaerobic conditions. During storage and when manure has been applied to the land, nitrous oxide is also produced as a byproduct of the denitrification process. Nitrous oxide (N2O) is 320 times more aggressive as a greenhouse gas than carbon dioxide and methane 25 times more than carbon dioxide.

By converting cow manure into methane biogas via anaerobic digestion, the millions of cattle in the United States would be able to produce 100 billion kilowatt hours of electricity, enough to power millions of homes across the United States. In fact, one cow can produce enough manure in one day to generate 3 kilowatt hours of electricity; only 2.4 kilowatt hours of electricity are needed to power a single 100-watt light bulb for one day. Furthermore, by converting cattle manure into methane biogas instead of letting it decompose, global warming gases could be reduced by 99 million metric tons or 4%.

Applications

A biogas bus in Linköping, Sweden

Biogas can be used for electricity production on sewage works, in a CHP gas engine, where the waste heat from the engine is conveniently used for heating the digester; cooking; space heating; water heating; and process heating. If compressed, it can replace compressed natural gas for use in vehicles, where it can fuel an internal combustion engine or fuel cells and is a much more effective displacer of carbon dioxide than the normal use in on-site CHP plants.

Biogas Upgrading

Raw biogas produced from digestion is roughly 60% methane and 29% CO_2 with trace elements of H_2S; it is not of high enough quality to be used as fuel gas for machinery. The corrosive nature of H_2S alone is enough to destroy the internals of a plant.

Methane in biogas can be concentrated via a biogas upgrader to the same standards as fossil natural gas, which itself has to go through a cleaning process, and becomes *biomethane*. If the local gas network allows, the producer of the biogas may use their distribution networks. Gas must be very clean to reach pipeline quality and must be of the correct composition for the distribution network to accept. Carbon dioxide, water, hydrogen sulfide, and particulates must be removed if present.

There are four main methods of upgrading: water washing, pressure swing adsorption, selexol adsorption, and amine gas treating. In addition to these, the use of membrane separation technology for biogas upgrading is increasing, and there are already several plants operating in Europe and USA.

The most prevalent method is water washing where high pressure gas flows into a column where the carbon dioxide and other trace elements are scrubbed by cascading water running counter-flow to the gas. This arrangement could deliver 98% methane with manufacturers guaranteeing maximum 2% methane loss in the system. It takes roughly between 3% and 6% of the total energy output in gas to run a biogas upgrading system.

Biogas Gas-grid Injection

Gas-grid injection is the injection of biogas into the methane grid (natural gas grid). Injections includes biogas until the breakthrough of micro combined heat and power two-thirds of all the energy produced by biogas power plants was lost (the heat), using the grid to transport the gas to customers, the electricity and the heat can be used for on-site generation resulting in a reduction of losses in the transportation of energy. Typical energy losses in natural gas transmission systems range from 1% to 2%. The current energy losses on a large electrical system range from 5% to 8%.

Biogas in Transport

"Biogaståget Amanda" ("The Biogas Train Amanda") train near Linköping station, Sweden

If concentrated and compressed, it can be used in vehicle transportation. Compressed biogas is becoming widely used in Sweden, Switzerland, and Germany. A biogas-powered train, named Biogaståget Amanda (The Biogas Train Amanda), has been in service in Sweden since 2005. Biogas powers automobiles. In 1974, a British documentary film titled *Sweet as a Nut* detailed the biogas production process from pig manure and showed how it fueled a custom-adapted combustion engine. In 2007, an estimated 12,000 vehicles were being fueled with upgraded biogas worldwide, mostly in Europe.

Measuring in Biogas Environments

Biogas is part of the wet gas and condensing gas (or air) category that includes mist or fog in the gas stream. The mist or fog is predominately water vapor that condenses on the sides of pipes or stacks throughout the gas flow. Biogas environments include wastewater digesters, landfills, and animal feeding operations (covered livestock lagoons).

Ultrasonic flow meters are one of the few devices capable of measuring in a biogas atmosphere. Most thermal flow meters are unable to provide reliable data because the moisture causes steady high flow readings and continuous flow spiking, although there are single-point insertion thermal mass flow meters capable of accurately monitoring biogas flows with minimal pressure drop. They can handle moisture variations that occur in the flow stream because of daily and seasonal temperature fluctuations, and account for the moisture in the flow stream to produce a dry gas value.

Legislation

European Union

The European Union has legislation regarding waste management and landfill sites called the Landfill Directive.

Countries such as the United Kingdom and Germany now have legislation in force that provides farmers with long-term revenue and energy security.

United States

The United States legislates against landfill gas as it contains VOCs. The United States Clean Air Act and Title 40 of the Code of Federal Regulations (CFR) requires landfill owners to estimate the quantity of non-methane organic compounds (NMOCs) emitted. If the estimated NMOC emissions exceeds 50 tonnes per year, the landfill owner is required to collect the gas and treat it to remove the entrained NMOCs. Treatment of the landfill gas is usually by combustion. Because of the remoteness of landfill sites, it is sometimes not economically feasible to produce electricity from the gas.

Global Developments

United States

With the many benefits of biogas, it is starting to become a popular source of energy and is starting to be used in the United States more. In 2003, the United States consumed 147 trillion BTU of energy from "landfill gas", about 0.6% of the total U.S. natural gas consumption. Methane biogas derived from cow manure is being tested in the U.S. According to a 2008 study, collected by the *Science and Children* magazine, methane biogas from cow manure would be sufficient to produce 100 billion kilowatt hours enough to power millions of homes across America. Furthermore, methane biogas has been tested to prove that it can reduce 99 million metric tons of greenhouse gas emissions or about 4% of the greenhouse gases produced by the United States.

In Vermont, for example, biogas generated on dairy farms was included in the CVPS Cow Power program. The program was originally offered by Central Vermont Public Service Corporation as a voluntary tariff and now with a recent merger with Green Mountain Power is now the GMP Cow Power Program. Customers can elect to pay a premium on their electric bill, and that premium is passed directly to the farms in the program. In Sheldon, Vermont, Green Mountain Dairy has provided renewable energy as part of the Cow Power program. It started when the brothers who own the farm, Bill and Brian Rowell, wanted to address some of the manure management challenges faced by dairy farms, including manure odor, and nutrient availability for the crops they need to grow to feed the animals. They installed an anaerobic digester to process the cow and milking center waste from their 950 cows to produce renewable energy, a bedding to replace sawdust, and a plant-friendly fertilizer. The energy and environmental attributes are sold to the GMP Cow Power program. On average, the system run by the Rowells produces enough electricity to power 300 to 350 other homes. The generator capacity is about 300 kilowatts.

In Hereford, Texas, cow manure is being used to power an ethanol power plant. By switching to methane biogas, the ethanol power plant has saved 1000 barrels of oil a day. Over all, the power plant has reduced transportation costs and will be opening many more jobs for future power plants that will rely on biogas.

In Oakley, Kansas, an ethanol plant considered to be one of the largest biogas facilities in North America is using Integrated Manure Utilization System "IMUS" to produce heat for its boilers by utilizing feedlot manure, municipal organics and ethanol plant waste. At full capacity the plant is expected to replace 90% of the fossil fuel used in the manufacturing process of ethanol.

Europe

The level of development varies greatly in Europe. While countries such as Germany, Austria and Sweden are fairly advanced in their use of biogas, there is a vast potential for this renewable energy source in the rest of the continent, especially in Eastern Europe. Different legal frameworks, education schemes and the availability of technology are among the prime reasons behind this untapped potential. Another challenge for the further progression of biogas has been negative public perception.

Initiated by the events of the gas crisis in Europe during December 2008, it was decided to launch the EU project "SEBE" (Sustainable and Innovative European Biogas Environment) which is financed under the CENTRAL programme. The goal is to address the energy dependence of Europe by establishing an online platform to combine knowledge and launch pilot projects aimed at raising awareness among the public and developing new biogas technologies.

In February 2009, the European Biogas Association (EBA) was founded in Brussels as a non-profit organisation to promote the deployment of sustainable biogas production and use in Europe. EBA's strategy defines three priorities: establish biogas as an important part of Europe's energy mix, promote source separation of household waste to increase the gas potential, and support the production of biomethane as vehicle fuel. In July 2013, it had 60 members from 24 countries across Europe.

UK

As of September 2013, there are about 130 non-sewage biogas plants in the UK. Most are on-farm, and some larger facilities exist off-farm, which are taking food and consumer wastes.

On 5 October 2010, biogas was injected into the UK gas grid for the first time. Sewage from over 30,000 Oxfordshire homes is sent to Didcot sewage treatment works, where it is treated in an anaerobic digestor to produce biogas, which is then cleaned to provide gas for approximately 200 homes.

In 2015 the Green-Energy company Ecotricity announced their plans to build three grid-injecting digester's.

Germany

Germany is Europe's biggest biogas producer and the market leader in biogas technology. In 2010 there were 5,905 biogas plants operating throughout the country: Lower Saxony, Bavaria, and the eastern federal states are the main regions. Most of these plants are employed as power plants. Usually the biogas plants are directly connected with a CHP which produces electric power by burning the bio methane. The electrical power is then fed into the public power grid. In 2010, the total installed electrical capacity of these power plants was 2,291 MW. The electricity supply was approximately 12.8 TWh, which is 12.6% of the total generated renewable electricity.

Biogas in Germany is primarily extracted by the co-fermentation of energy crops (called 'NawaRo', an abbreviation of *nachwachsende Rohstoffe*, German for renewable resources) mixed with manure. The main crop used is corn. Organic waste and industrial and agricultural residues such as waste from the food industry are also used for biogas generation.In this respect, biogas production in Germany differs significantly from the UK, where biogas generated from landfill sites is most common.

Biogas production in Germany has developed rapidly over the last 20 years. The main reason is the legally created frameworks. Government support of renewable energy started in 1991 with the Electricity Feed-in Act (*StrEG*). This law guaranteed the producers of energy from renewable sources the feed into the public power grid, thus the power companies were forced to take all produced energy from independent private producers of green energy. In 2000 the Electricity Feed-in Act was replaced by the Renewable Energy Sources Act (*EEG*). This law even guaranteed a fixed compensation for the produced electric power over 20 years. The amount of around 8 ¢/kWh gave farmers the opportunity to become energy suppliers and gain a further source of income.

The German agricultural biogas production was given a further push in 2004 by implementing the so-called NawaRo-Bonus. This is a special payment given for the use of renewable resources, that is, energy crops. In 2007 the German government stressed its intention to invest further effort and support in improving the renewable energy supply to provide an answer on growing climate challenges and increasing oil prices by the 'Integrated Climate and Energy Programme'.

This continual trend of renewable energy promotion induces a number of challenges facing the management and organisation of renewable energy supply that has also several impacts on the biogas production. The first challenge to be noticed is the high area-consuming of the biogas electric power supply. In 2011 energy crops for biogas production consumed an area of circa 800,000 ha in Germany. This high demand of agricultural areas generates new competitions with the food industries that did not exist hitherto. Moreover, new industries and markets were created in predominately rural regions entailing different new players with an economic, political and civil background. Their influence and acting has to be governed to gain all advantages this new source of energy is offering. Finally biogas will furthermore play an important role in the German renewable energy supply if good governance is focused.

Indian Subcontinent

Biogas in India has been traditionally based on dairy manure as feed stock and these "gobar" gas plants have been in operation for a long period of time, especially in rural India. In the last 2-3 decades, research organisations with a focus on rural energy security have enhanced the design of the systems resulting in newer efficient low cost designs such as the Deenabandhu model.

The Deenabandhu Model is a new biogas-production model popular in India. (*Deenabandhu* means "friend of the helpless.") The unit usually has a capacity of 2 to 3 cubic metres. It is constructed using bricks or by a ferrocement mixture. In India, the brick model costs slightly more than the ferrocement model; however, India's Ministry of New and Renewable Energy offers some subsidy per model constructed.

LPG (Liquefied Petroleum Gas) is a key source of cooking fuel in urban India and its prices have been increasing along with the global fuel prices. Also the heavy subsidies provided by the successive governments in promoting LPG as a domestic cooking fuel has become a financial burden renewing the focus on biogas as a cooking fuel alternative in urban establishments. This has led to the development of prefabricated digester for modular deployments as compared to RCC and cement structures which take a longer duration to construct. Renewed focus on process technology like the Biourja process model has enhanced the stature of medium and large scale anaerobic digester in India as a potential alternative to LPG as primary cooking fuel.

In India, Nepal, Pakistan and Bangladesh biogas produced from the anaerobic digestion of manure in small-scale digestion facilities is called gobar gas; it is estimated that such facilities exist in over 2 million households in India, 50,000 in Bangladesh and thousands in Pakistan, particularly North Punjab, due to the thriving population of livestock. The digester is an airtight circular pit made of concrete with a pipe connection. The manure is directed to the pit, usually straight from the cattle shed. The pit is filled with a required quantity of wastewater. The gas pipe is connected to the kitchen fireplace through control valves. The combustion of this biogas has very little odour or smoke. Owing to simplicity in implementation and use of cheap raw materials in villages, it is one of the most environmentally sound energy sources for rural needs. One type of these system is the Sintex Digester. Some designs use vermiculture to further enhance the slurry produced by the biogas plant for use as compost.

To create awareness and associate the people interested in biogas, the Indian Biogas Association was formed. It aspires to be a unique blend of nationwide operators, manufacturers and planners of biogas plants, and representatives from science and research. The association was founded in 2010 and is now ready to start mushrooming. Its motto is "propagating Biogas in a sustainable way".

In Pakistan, the Rural Support Programmes Network is running the Pakistan Domestic Biogas Programme which has installed 5,360 biogas plants and has trained in excess of 200 masons on the technology and aims to develop the Biogas Sector in Pakistan.

In Nepal, the government provides subsidies to build biogas plant.

China

The Chinese had experimented the applications of biogas since 1958. Around 1970, China had installed 6,000,000 digesters in an effort to make agriculture more efficient. During the last years the technology has met high growth rates. This seems to be the earliest developments in generating biogas from agricultural waste.

In Developing Nations

Domestic biogas plants convert livestock manure and night soil into biogas and slurry, the fermented manure. This technology is feasible for small-holders with livestock producing 50 kg manure per day, an equivalent of about 6 pigs or 3 cows. This manure has to be collectable to mix it with water and feed it into the plant. Toilets can be connected. Another precondition is the temperature that affects the fermentation process. With an optimum at 36 C° the technology especially applies for those living in a (sub) tropical climate. This makes the technology for small holders in developing countries often suitable.

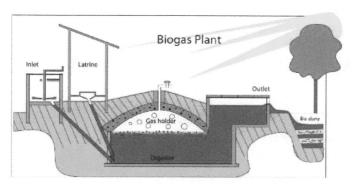

Simple sketch of household biogas plant

Depending on size and location, a typical brick made fixed dome biogas plant can be installed at the yard of a rural household with the investment between US$300 to $500 in Asian countries and up to $1400 in the African context. A high quality biogas plant needs minimum maintenance costs and can produce gas for at least 15–20 years without major problems and re-investments. For the user, biogas provides clean cooking energy, reduces indoor air pollution, and reduces the time needed for traditional biomass collection, especially for women and children. The slurry is a clean organic fertilizer that potentially increases agricultural productivity.

Domestic biogas technology is a proven and established technology in many parts of the world, especially Asia. Several countries in this region have embarked on large-scale programmes on domestic biogas, such as China and India.

The Netherlands Development Organisation, SNV, supports national programmes on domestic biogas that aim to establish commercial-viable domestic biogas sectors in which local companies market, install and service biogas plants for households. In Asia, SNV is working in Nepal, Vietnam, Bangladesh, Bhutan, Cambodia, Lao PDR, Pakistan and Indonesia, and in Africa; Rwanda, Senegal, Burkina Faso, Ethiopia, Tanzania, Uganda, Kenya, Benin and Cameroon.

In South Africa a prebuilt Biogas system is manufactured and sold. One key feature is that installation requires less skill and is quicker to install as the digester tank is premade plastic.

Society and Culture

In the 1985 Australian film *Mad Max Beyond Thunderdome* the post-apocalyptic settlement Barter town is powered by a central biogas system based upon a piggery. As well as providing electricity, methane is used to power Barter's vehicles.

"Cow Town", written in the early 1940s, discuss the travails of a city vastly built on cow manure and the hardships brought upon by the resulting methane biogas. Carter McCormick, an engineer from a town outside the city, is sent in to figure out a way to utilize this gas to help power, rather than suffocate, the city.

References

- J. Goettemoeller; A. Goettemoeller (2007). Sustainable Ethanol: Biofuels, Biorefineries, Cellulosic Biomass, Flex-Fuel Vehicles, and Sustainable Farming for Energy Independence. Prairie Oak Publishing, Maryville, Missouri. p. 42. ISBN 978-0-9786293-0-4.

- Borowitzka, M. A. (2013). "Energy from Microalgae: A Short History". Algae for Biofuels and Energy. p. 1. doi:10.1007/978-94-007-5479-9_1. ISBN 978-94-007-5478-2.

- Voelcker, John (2016-06-14). "Nissan takes a different approach to fuel cells: ethanol". Green Car Reports. Retrieved 2016-06-16.

- "Biogas Flows Through Germany's Grid Big Time - Renewable Energy News Article". 14 March 2012. Archived from the original on 14 March 2012. Retrieved 17 June 2016.

- "World Energy Outlook 2006" (PDF). Worldenergyoutlook.org. Archived from the original (PDF) on 28 September 2007. Retrieved 20 January 2015.

- "Ethanol Producer Magazine – The Latest News and Data About Ethanol Production". Ethanolproducer.com. Retrieved 20 January 2015.

- "01.26.2006 - Ethanol can replace gasoline with significant energy savings, comparable impact on greenhouse gases". Berkeley.edu. Retrieved 20 January 2015.

- "Catalytic deoxygenation of microalgae oil to green hydrocarbons - Green Chemistry (RSC Publishing)". pubs.rsc.org. Retrieved 2015-06-08.

- Administrator. "Biogas CHP - Alfagy - Profitable Greener Energy via CHP, Cogen and Biomass Boiler using Wood, Biogas, Natural Gas, Biodiesel, Vegetable Oil, Syngas and Straw". Retrieved 15 May 2015.

- Cohn, D.R.; Bromberg, L.; Heywood, J.B. (April 20, 2005), "Direct Injection Ethanol Boosted Gasoline Engines: Biofuel Leveraging for Cost Effective Reduction of Oil Dependence and CO_2 Emissions. MIT Report PSFC/JA-06-16" (PDF), MIT Energy Initiative, Cambridge, MA: MIT Plasma Science and Fusion Center, retrieved November 23, 2014

- Jim Lane (2013-08-01). "INEOS Bio produces cellulosic ethanol from waste, at commercial scale – print-friendly". Biofuels Digest. Retrieved 2014-06-15.

- "National Algal Biofuels Technology Roadmap" (PDF). US Department of Energy, Office of Energy Efficiency and Renewable Energy, Biomass Program. Retrieved 3 April 2014.

- "Effects of nitrogen on growth and carbohydrate formation in Porphyridium cruentum". Open Life Sciences. 9. 2013-09-21. doi:10.2478/s11535-013-0248-z. Retrieved 2014-08-19.

- "Growth Rates of Emission-Fed Algae Show Viability of New Biomass Crop" (PDF). Arizona Public Service Company (APS) and GreenFuel Technologies Corporation (GFT). 26 September 2008. Archived from the original (PDF) on 2008-05-21. Retrieved 15 December 2013.

- Bullis, Kevin (2007-02-05). "Algae-Based Fuels Set to Bloom | MIT Technology Review". Technologyreview.com. Retrieved 29 November 2013.

Aviation Biofuel: A Comprehensive Study

Biofuels are seen as a means for replacing all of human energy needs from domestic fuel consumption to vehicular fuel. It can be the cause of reduction of greenhouse gases caused by the aircrafts and the aviation industry. The biofuel discussed in this section is sustainable aviation fuel. The focus of this chapter is the relevancy of biofuels in the aviation industry.

Aviation Biofuel

Aviation biofuel is a biofuel used for aircraft. It is considered by some to be the primary means by which the aviation industry can reduce its carbon footprint. After a multi-year technical review from aircraft makers, engine manufacturers and oil companies, biofuels were approved for commercial use in July 2011. Since then, some airlines have experimented with using of biofuels on commercial flights. The focus of the industry has now turned to second generation sustainable biofuels (sustainable aviation fuels) that do not compete with food supplies nor are major consumers of prime agricultural land or fresh water.

US Marine AV-8B Harrier test flight using a 50-50 biofuel blend in 2011.

Rationale for Aviation Biofuels

Aviation's share of the greenhouse gas emissions is poised to grow, as air travel increases and ground vehicles use more alternative fuels like ethanol and biodiesel. Currently aviation represents 2% of global emissions, but is expected to grow to 3% by 2050. In addition to building more fuel efficient aircraft and operating them more efficiently, changing the fuel source is one of the few options the aviation industry has for reducing its carbon footprint. While solar, electric and hydrogen propelled aircraft are being researched, it is not expected they will be feasible in the near or medium term due to aviation's need for high power-to-weight ratio and globally compatible infrastructure.

Concerns and Challenges

Biodiesel that is stored for long periods of time is more likely to oxidize, especially at low temperatures, causing it to gel. Some additives improve the cold weather tolerance of biodiesel, but only by a few degrees. Nitrile-based rubber materials expand in the presence of aromatic compounds found in conventional petroleum fuel. Pure biofuels that aren't mixed with petroleum and don't contain paraffin-based additives may cause rubber seals and hoses to shrink. Manufacturers are starting to use a synthetic rubber substitute called Viton for seals and hoses. Viton isn't adversely affected by biofuels.

Industry Commitments and Collaborations

The International Air Transport Association (IATA) supports research, development and deployment of alternative fuels. IATA thinks a 6% share of sustainable 2nd generation biofuels is achievable by 2020, and Boeing supports a target of 1% of global aviation fuels by 2015. This is in support of the goals of the aviation industry reaching carbon neutral growth by 2020 and a 50% decrease in carbon emissions by 2050 (relative to a 2005 baseline)

A group of interested airlines has formed the Sustainable Aviation Fuel Users Group (SAFUG). The group was formed in 2008 in cooperation with support from NGOs such as Natural Resources Defense Council and The Roundtable For Sustainable Biofuels (RSB). Member airlines represent more than 15% of the industry, and all member CEOs have signed a pledge to work on the development and use of sustainable biofuels for aviation.

Boeing is joining other aviation-related members in the Algal Biomass Organization (ABO).

Production Routes and Sources

Jet fuel is a mixture of a large number of different hydrocarbons. The range of their sizes (molecular weights or carbon numbers) is restricted by the requirements for the product, for example, freezing point or smoke point. Jet fuels are sometimes classified as kerosene or naphtha-type. Kerosene-type fuels include Jet A, Jet A-1, JP-5 and JP-8. Naphtha-type jet fuels, sometimes referred to as "wide-cut" jet fuel, include Jet B and JP-4.

"Drop-in" biofuels are biofuels that are completely interchangeable with conventional fuels. Deriving "drop-in" jet fuel from bio-based sources is ASTM approved via two routes.

Bio-SPK

The first route involves using oil which is extracted from plant sources like jatropha, algae, tallows, other waste oils, Babassu and camelina to produce bio-SPK (Bio derived synthetic paraffinic Kerosene) by cracking and hydroprocessing.

The growing of algae to make jet fuel is a promising but still emerging technology. Companies working on algae jet fuel are Solazyme, Honeywell UOP, Solena, Sapphire Energy, Imperium Renewables, and Aquaflow Bionomic Corporation. Universities working on algae jet fuel are Arizona State University and Cranfield University

Major investors for algae based SPK research are Boeing, Honeywell/UOP, Air New Zealand, Continental Airlines, Japan Airlines, and General Electric.

FT-SPK

The second route involves processing solid biomass using pyrolysis to produce pyrolysis oil or gasification to produce a syngas which is then processed into FT SPK (Fischer–Tropsch Synthetic Paraffinic Kerosene).

Future Production Routes

Further research is being done on an alcohol-to-jet pathway where alcohols such as ethanol or butanol are de-oxygenated and processed into jet fuels. In addition, routes that use synthetic biology to directly create hydro-carbons are being researched.

Commercial and Demonstration Flights

Since 2008, a large number of test flights have been conducted, and since ASTM approval in July 2011, several commercial flights with passengers have also occurred.

Demonstration Flights

Date	Operator	Platform	Biofuel	Notes
2 October 2007	GreenFlight International	Aero L-29 Delfin	Waste Vegetable Oil	Greenflight International made the very first flight of an aircraft powered entirely by 100% biofuel from the Reno, Stead airport on the afternoon of 2 October 2007. There is no citation for this entry - it was made by the pilot that flew it. In November 2008 the same aircraft and flight crew flew from Reno, NV to Leesburg, FL using 100% biofuel for the first seven of the nine legs, the remaining three were completed on a 50% biofuel 50% JetA blend.
February 2008	Virgin Atlantic	Boeing 747	Coconut and Babassu	Virgin flew a biofuel test flight between London and Amsterdam, using a 20% blend of biofuels in one of its engines
December 2008	Air New Zealand	Boeing 747	Jatropha	A two-hour test flight using a 50-50 mixture of the new biofuel with Jet A-1 in the number one position Rolls Royce RB-211 engine of 747-400 ZK-NBS, was successfully completed on 30 December 2008. The engine was then removed to be scrutinised and studied to identify any differences between the Jatropha blend and regular Jet A1. No effects to performance were found.

January 2009	Continental Airlines	Boeing 737	Algae and jatropha	Continental Airlines ran the first flight of an algae-fueled jet. The flight from Houston's George Bush Intercontinental Airport completed a circuit over the Gulf of Mexico. The pilots on board executed a series of tests at 38,000 feet (12,000 m), including a mid-flight engine shutdown. Larry Kellner, chief executive of Continental Airlines, said they had tested a drop-in fuel which meant that no modification to the engine was required. The fuel was praised for having a low flash point and sufficiently low freezing point, issues that have been problematic for other bio-fuels.
January 2009	Japan Airlines	Boeing 747	Camelina, jatropha and algae	Japan Airlines conducted a one and a half hour flight with one engine burning a 50/50 mix of Jet-A and biofuel from the *Camelina* plant.
April 2010	US Navy	F/A-18	Camelina	The Navy tested this biofuel blend on the F/A-18 Super Hornet aka "Green Hornet". Results from those tests indicated the aircraft performed as expected through its full flight envelope with no degradation of capability.
March 2010	US Air Force	A-10	Camelina	On March 25, 2010, the United States Air Force conducted the first flight of an aircraft with all engines powered by a biofuel blend. The flight, performed on an A-10 at Eglin Air Force Base, used a 50/50 blend of JP-8 and Camelina-based fuel.
June 2010	Dutch Military	Ah-64 Apache Helicopter	Waste cooking oil	
June 2010	EADS	Diamond D42	Algae	Occurred at an air show in Berlin in June 2010.
November 2010	US Navy	MH-60S Seahawk	Camelina	Flown on 50/50 biofuel blend Nov. 18, 2010 in Patuxent River, Md. The helicopter, from Air Test and Evaluation Squadron 21 at Naval Air Station Patuxent River tested a fuel mixture made from the Camelina seed.
November 2010	TAM	Airbus 320	Jatropha	A 50/50 biofuel blend of conventional and jatropha oil
June 2011	Boeing	Boeing 747-8F	Camelina	Boeing flew its new model 747-8F to the Paris Air Show with all four engines burning a 15% mix of biofuel from camelina
June 2011	Honeywell	Gulfstream G450	Camelina	The first transatlantic biofuels flight using a 50/50 blend of camelina-based biofuel and petroleum-based fuel.
August 2011	US Navy	T-45	Camelina	Successfully flew a T-45 training aircraft using biofuels at the Naval Air Station (NAS) in Patuxent River, Maryland. The flight was completed by the "Salty Dogs" of Air Test and Evaluation Squadron 23 flying on biofuel mixture of 50/50 petroleum-based JP-5 jet fuel and plant-based camelina.
September 2011	US Navy	AV-8B	Camelina	Naval Air Warfare Center Weapons Division, China Lake performed the first bio-fuel flight test in AV-8B Harrier from Air Test and Evaluation Squadron 31.
October 2011	Air China	Boeing 747-400	Jatropha	Air China flew China's first flight using aviation biofuels. The flight was conducted using Chinese grown jatropha oil from PetroChina. The flight was 2 hours in duration above Beijing, and used 50% biofuel in 1 engine.
November 2011	Continental Airlines	Boeing 737-800	Algae	United / Continental flew a biofuel flight from IAH to ORD on algae jet fuel supplied by Solazyme. The fuel was partially derived from genetically modified algae that feed on plant waste and produce oil. It was the first biofuel-powered air service in the US.

November 2011	Alaska Airlines	Boeing 737 and Bombardier Q400	Algae	Alaska Airlines and its sister carrier, Horizon Air, converted 75 flights on their schedules to run on a fuel mixture of 80% kerosene and 20% biofuel derived from used cooking oil. The biofuel was made by Dynamic Fuels, a joint venture of Tyson Foods and Syntroleum Corp.
January 2012	Etihad Airways	Boeing 777-300ER	vegetable cooking oil	Etihad Airways conducted a biofuel flight from Abu Dhabi to Seattle using a combination of traditional jet fuel and fuel based on recycled vegetable cooking oil
April 2012	Qantas Airways	Airbus A330	Refined cooking oil	Qantas Airways used 50/50 mix of biofuel supplied by SkyNRG and Jet-A fuel in one engine for a flight from Sydney to Adelaide.
April 2012	Porter Airlines	Bombardier Q400	*Camelina sativa* and *Brassica carinata*	Porter Airlines used 50/50 mix of biofuel (49% *Camelina sativa* and 1% *Brassica carinataand*) and Jet-A fuel in one engine for a flight from Toronto to Ottawa.
October 2012	NRC	Dassault Falcon 20	Carinata	First jet to fly on 100% biofuels that meet petroleum specifications without blending. Fuel was produced by Applied Research Associates (ARA) and Chevron Lummus Global (CLG) from carinata oil supplied by Agrisoma Biosciences.
March 2013	Paramus Flying Club	Cessna 182	Waste cooking oil	First piston engine aircraft to fly with a 50/50 blend of aviation biofuel and conventional Jet-A (as specified by ASTM D7566). First piston engine aircraft to fly with a biofuel blend operating under a standard (not experimental) airworthiness certificate. Demonstration flight from North Central State Airport (KSFZ) in Rhode Island to First Flight Airport (KFFA) in North Carolina took place on March 2, 2013. The Cessna 182 had been converted under STC to be powered by an SMA jet-fuel diesel cycle piston engine, and the blended biofuel was provided by SkyNRG of Holland.

Commercial Flights

Date	Operator	Platform	Biofuel	Notes
Jun 2011	KLM	Boeing 737-800	Used cooking oil	KLM flew the world's first commercial biofuel flight, carrying 171 passengers from Amsterdam to Paris
Jul 2011	Lufthansa	Airbus A321	Jatropha, camelina plants and animal fats	First German commercial biofuel's flight, and the start of 6 month regular series of flights from Hamburg to Frankfurt with one of the two engines use biofuel. It officially ended on January 12, 2012 with a flight from Frankfurt to Washington and would not take biofuel further unless the biofuel was more widely produced.
Jul 2011	Finnair	Airbus A319	Used cooking oil	The 1,500 km journey between Amsterdam and Helsinki was fuelled with a mix of 50 per cent biofuel derived from used cooking oil and 50 per cent conventional jet fuel. Finnair says it will conduct at least three weekly Amsterdam-to-Helsinki flights using the biofuel blend in both of the aircraft's engines. Refueling will be done at Amsterdam Airport Schiphol.
Jul 2011	Interjet	Airbus A320	Jatropha	Flight was powered by 27% jatropha between Mexico City and Tuxtla Gutierrez
Aug 2011	AeroMexico	Boeing 777-200	Jatropha	Aeromexico flew the world's first trans-Atlantic revenue flight, from Mexico City to Madrid with passengers

Oct 2011	Thomson Airways	Boeing 757-200	Used cooking oil	Thomson flew the UK's first commercial biofuel flight from Birmingham Airport on one engine using biofuel from used cooking oil, supplied by SkyNRG
November 2011	Continental Airlines	Boeing 737-800	Algae	United / Continental flew biofuel flight from IAH to ORD on algae jet fuel, which supplied by Solazyme
April 19, 2012	Jetstar Airways	Boeing 777-206ER	Refined cooking oil	JQ flight 705 departed Melbourne at 0950 and arrived in Hobart at 1105 supplied by SkyNRG
March 13, 2013	KLM	Boeing 777-206ER	Used cooking oil	KLM begins weekly flights by a Boeing 777-200 between John F. Kennedy Airport in New York City, USA and Amsterdam's Schiphol Airport, Netherlands using Biofuel supplied by SkyNRG
May 16, 2014	KLM	Airbus A330-200	Used cooking oil	KLM begins weekly flights by an Airbus A330-200 between Queen Beatrix International Airport, in Oranjestad Aruba and Amsterdam's Schiphol Airport, Netherlands (with a stop-over in Bonaire) using Biofuel supplied by SkyNRG
Aug 4, 2014	Gol Transportes Aéreos	Boeing 737-700	Inedible corn oil and used cooking oil	Gol Flight 2152 took off from Rio Santos Dumont Airport (SDU) towards Brasilia (BSB) with a 4% mix of bio jetfuel
Nov 7, 2014	Scandinavian Airlines	Boeing 737-600	Used cooking oil	SAS Flight SK2064 flew their first ever flight using biofuel between Stockholm and Östersund using a 10% blend of JET A1 based on used cooking oil. It was also the first flight from Arlanda Airport
Nov 11, 2014	Scandinavian Airlines	Boeing 737-700	Used cooking oil	SAS Flight SK371 flew the first ever Norwegian domestic flight using bio-fuel between Trondheim and Oslo using a 48% blend of JET A1 based on used cooking oil
Mar 21, 2015	Hainan Airlines	Boeing 737-800	Used cooking oil	Hainan Airlines conducted China's first commercial biofuel flight carrying 156 passengers from Shanghai to Beijing. The fuel, supplied by Sinopec, was a fuel blend of approximately 50 percent aviation biofuel mixed with conventional petroleum jet fuel.

Environmental Effects

A life cycle assessment by the Yale School of Forestry on jatropha, one source of potential biofuels, estimated using it could reduce greenhouse gas emissions by up to 85% if former agro-pastoral land is used, or increase emissions by up to 60% if natural woodland is converted to use. In addition, biofuels do not contain sulfur compounds and thus do not emit sulfur dioxide.

Many different standards exist for certification of sustainable biofuels. One such standard often cited by airlines is the one developed by the The Roundtable For Sustainable Biofuels. Nearly all such standards include a minimum amount of greenhouse gas reduction and consideration that biofuels do not compete with food.

Sustainable Aviation Fuel

Sustainable aviation fuel (SAF) is the name given to advanced aviation biofuel types used in jet aircraft and certified as being sustainable by a reputable independent third-party, such as the

Roundtable on Sustainable Biomaterials (RSB). This certification is in addition to the safety and performance certification, issued by global standards body ASTM International, that all jet fuel is required to meet in order to be approved for use in regular passenger flights.

Certification

A SAF sustainability certification verifies that the fuel product, mainly focussing on the biomass feedstock, has met criteria focussed around long-term global environmental, social and economic "triple-bottom-line" sustainability considerations. Under many carbon emission regulation schemes, such as the European Union Emissions Trading Scheme, a certified SAF product may be granted an exemption from an associated carbon compliance liability cost. This marginally improves the economic competitiveness of environmentally favourable SAF over traditional fossil-based jet fuel. However, in the near term there are several commercialisation and regulatory hurdles that are yet to be overcome through the collaboration of a variety of stakeholders for SAF products to meet price parity with traditional jet fuel and to enable widespread uptake.

The first reputable body to launch a sustainable biofuel certification system applicable to SAF was the academic European-based Roundtable on Sustainable Biomaterials (RSB) NGO. This multi-stakeholder organization set a global benchmark standard on which the sustainability integrity of advanced aviation biofuel types seeking to use the claim of being a Sustainable Aviation Fuel can be judged. Leading airlines in the aviation industry and other signatories to the Sustainable Aviation Fuel Users Group pledge support the RSB as the preferred provider of SAF certification. These airlines believe it important for any proposed aviation biofuels have independently certified sustainable biofuel long term environmental benefits compared to the status quo in order to ensure their successful uptake and marketability

Global Impact

As emissions trading schemes and other carbon compliance regimes are emerging globally certain biofuels are likely to be exempt, "zero rated", by governments from having an associated carbon compliance liability due to their closed-emissions-loop renewable nature if they can also prove their wider sustainability credentials. For example, in the European Union Emissions Trading Scheme it has been proposed by SAFUG that only aviation biofuels that have been certified as sustainable by the RSB or similar bodies would be zero rated. This proposal has been accepted.

SAFUG was formed by a group of interested airlines in 2008 under the auspices of Boeing Commercial Airplanes and in cooperation with support from NGOs such as Natural Resources Defense Council. Member airlines represent more than 15% of the industry, and all member CEOs have signed a pledge to work on the development and use of Sustainable Aviation Fuel.

In addition to SAF certification, the integrity of aviation biofuel producers and their product can be assessed by further means such as by using Richard Branson's Carbon War Room Renewable Jets Fuels initiative. A leading independent NGO focused on this issue is the Sustainable Sky Institute

References

- This story was written Robert Kaper, Naval Air Systems Command Public Affairs. "Fuels Team Plans Super Hornet Biofuels Flight Test". Retrieved 25 September 2015.

- "Finnair's scheduled commercial biofuel flight marks a step towards more sustainable flying, says airline on GreenAir Online". Retrieved 25 September 2015.

- Eric Loveday (20 July 2011). "Finnair to attempt longest commercial biofuel flight in aviation history". Autoblog. Retrieved 25 September 2015.

- "Precisamos de bioindústria funcionando com excelência no país". Ubrabio - União Brasileira de Biodiesel e Bioquerosene. Retrieved 25 September 2015.

- "RSB Roundtable on Sustainable Biomaterials | Roundtable on Sustainable Biomaterials" (PDF). Rsb.epfl.ch. 2013-10-17. Retrieved 2013-10-24.

- "Renewable and Alternative Energy Fact Sheet" (PDF). Agricultural Research and Cooperative Extension. Penn State College of Agricultural Science. Retrieved March 7, 2012.

- "Technical Report: Near-Term Feasibility of Alternative Jet Fuels" (PDF). Sponsored by the FAA. Authored by MIT staff. Published by RAND Corporation. Retrieved August 22, 2012.

- Dunn, Graham (24 February 2008). "Partners carry out first biofuel flight using Virgin 747". Flight International. Retrieved 25 August 2012.

- Cattermole, Tannith (June 26, 2011). "Gulfstream G450 crosses the Atlantic on 50/50 biofuel-jetfuel blend". GizMag. Retrieved March 7, 2012.

- "NRC Flies World's First Civil Jet Powered by 100 Percent Biofuel". Aero-news Network. 7 November 2012. Retrieved 21 November 2012.

- "Alaska Airlines Launching Biofuel-Powered Commercial Service In The United States" (Press release). Alaska Airlines. November 7, 2011. Retrieved December 30, 2011.

- "Aviation Fuel Standard Takes Flight". Aviation Fuel Standard Takes Flight. ASTM. September–October 2011. Retrieved 1 April 2012.

Biomass used as a Biofuel

Biomass can be used as a source of energy, and is mostly referred to plants or plant-based materials that are used for the manufacture of various sources of fuel. The conversion of biomass into biofuel can be achieved by different methods which are broadly classified in this chapter. Biofuels are best understood in confluence with the major topics listed in the following chapter.

Biomass

Sugarcane plantation in Brazil. Sugarcane bagasse is a type of biomass.

Biomass is organic matter derived from living, or recently living organisms. Biomass can be used as a source of energy and it most often refers to plants or plant-based materials which are not used for food or feed, and are specifically called lignocellulosic biomass. As an energy source, biomass can either be used directly via combustion to produce heat, or indirectly after converting it to various forms of biofuel. Conversion of biomass to biofuel can be achieved by different methods which are broadly classified into: *thermal*, *chemical*, and *biochemical* methods.

A cogeneration plant in Metz, France. The station uses waste wood biomass as an energy source, and provides electricity and heat for 30,000 dwellings.

Stump harvesting increases the recovery of biomass from a forest.

Biomass Sources

Eucalyptus in Brazil. Remains of the tree are reused for power generation.

Historically, humans have harnessed biomass-derived energy since the time when people began burning wood to make fire. Even today, biomass is the only source of fuel for domestic use in many developing countries. Biomass is all biologically-produced matter based in carbon, hydrogen and oxygen. The estimated biomass production in the world is 104.9 petagrams (104.9 * 10^{15} g - about 105 billion metric tons) of carbon per year, about half in the ocean and half on land.

Wood remains the largest biomass energy source today; examples include forest residues (such as dead trees, branches and tree stumps), yard clippings, wood chips and even municipal solid waste. Wood energy is derived by using lignocellulosic biomass (second-generation biofuels) as fuel. Harvested wood may be used directly as a fuel or collected from wood waste streams. The largest source of energy from wood is pulping liquor or "black liquor," a waste product from processes of the pulp, paper and paperboard industry. In the second sense, biomass includes plant or animal matter that can be converted into fibers or other industrial chemicals, including biofuels. Industrial biomass can be grown from numerous types of plants, including miscanthus, switchgrass, hemp, corn, poplar, willow, sorghum, sugarcane, bamboo, and a variety of tree species, ranging from eucalyptus to oil palm (palm oil).

Based on the source of biomass, biofuels are classified broadly into two major categories. First-generation biofuels are derived from sources such as sugarcane and corn starch. Sugars present in this biomass are fermented to produce bioethanol, an alcohol fuel which can be used directly in a fuel cell to produce electricity or serve as an additive to gasoline. However, utilizing food-based resources for fuel production only aggravates the food shortage problem. Second-generation biofuels, on the other hand, utilize non-food-based biomass sources such as agriculture and municipal waste. These biofuels mostly consist of lignocellulosic biomass, which is not edible and is a low-value waste for many industries. Despite being the favored alternative, economical production of second-generation biofuel is not yet achieved due to technological issues. These issues arise mainly due to chemical inertness and structural rigidity of lignocellulosic biomass.

Plant energy is produced by crops specifically grown for use as fuel that offer high biomass output per hectare with low input energy. Some examples of these plants are wheat, which typically yields 7.5–8 tonnes of grain per hectare, and straw, which typically yields 3.5–5 tonnes per hectare in the UK. The grain can be used for liquid transportation fuels while the straw can be burned to produce heat or electricity. Plant biomass can also be degraded from cellulose to glucose through a series of chemical treatments, and the resulting sugar can then be used as a first-generation biofuel.

The main contributors of waste energy are municipal solid waste, manufacturing waste, and landfill gas. Energy derived from biomass is projected to be the largest non-hydroelectric renewable resource of electricity in the US between 2000 and 2020.

Biomass can be converted to other usable forms of energy like methane gas or transportation fuels like ethanol and biodiesel. Rotting garbage, and agricultural and human waste, all release methane gas, also called landfill gas or biogas. Crops such as corn and sugarcane can be fermented to produce the transportation fuel ethanol. Biodiesel, another transportation fuel, can be produced from leftover food products like vegetable oils and animal fats. Also, biomass-to-liquids (called "BTLs") and cellulosic ethanol are still under research.

There is research involving algae, or algae-derived, biomass, as this non-food resource can be produced at rates five to ten times those of other types of land-based agriculture, such as corn and soy. Once harvested, it can be fermented to produce biofuels such as ethanol, butanol, and methane, as well as biodiesel and hydrogen. Efforts are being made to identify which species of algae are most suitable for energy production. Genetic engineering approaches could also be utilized to improve microalgae as a source of biofuel.

The biomass used for electricity generation varies by region. Forest by-products, such as wood residues, are common in the US. Agricultural waste is common in Mauritius (sugar cane residue) and Southeast Asia (rice husks). Animal husbandry residues, such as poultry litter, are common in the UK.

As of 2015, a new bioenergy sewage treatment process aimed at developing countries is under trial; the Omni Processor is a self-sustaining process which uses sewerage solids as fuel in a process to convert waste water into drinking water, with surplus electrical energy being generated for export.

Comparison of Total Plant Biomass Yields (Dry Basis)

World Resources

If the total annual primary production of biomass is just over 100 billion (1.0E+11) tonnes of Carbon /yr, and the energy reserve per metric tonne of biomass is between about 1.5E3 – 3E3 Kilowatt hours (5E6 – 10E6 BTU), or 24.8 TW average, then biomass could in principle provide 1.4 times the approximate annual 150E3 Terrawatt.hours required for the current world energy consumption. For reference, the total solar power on Earth is 174 kTW. The biomass equivalent to solar energy ratio is 143 ppm (parts per million), given current living system coverage on Earth. Best in class solar cell efficiency is (20-40)%. Additionally, Earth's internal radioactive energy production, largely the driver for volcanic activity, continental drift, etc., is in the same range of power, 20 TW. At some 50% carbon mass content in biomass, annual production, this corresponds to about 6% of atmospheric carbon content in CO_2 (for the current 400 ppm).

> (1.0E+11 tonnes biomass annually produced approximately 25 TW)
> Annual world biomass energy equivalent =16.7 - 33.4 TW.
> Annual world energy consumption =17.7. On average, biomass production is 1.4 times larger than world energy consumption.

Common Commodity Food Crops

- Agave: 1–21 tons/acre

- Alfalfa: 4–6 tons/acre

- Barley: grains – 1.6–2.8 tons/acre, straw – 0.9–2.5 tons/acre, total – 2.5–5.3 tons/acre

- Corn: grains – 3.2–4.9 tons/acre, stalks and stovers – 2.3–3.4 tons/acre, total – 5.5–8.3 tons/acre

- Jerusalem artichokes: tubers 1–8 tons/acre, tops 2–13 tons/acre, total 9–13 tons/acre

- Oats: grains – 1.4–5.4 tons/acre, straw – 1.9–3.2 tons/acre, total – 3.3–8.6 tons/acre

- Rye: grains – 2.1–2.4 tons/acre, straw – 2.4–3.4 tons/acre, total – 4.5–5.8 tons/acre
- Wheat: grains – 1.2–4.1 tons/acre, straw – 1.6–3.8 tons/acre, total – 2.8–7.9 tons/acre

Woody Crops

- Oil palm: fronds 11 ton/acre, whole fruit bunches 1 ton/acre, trunks 30 ton/acre

Not Yet in Commercial Planting

- Giant miscanthus: 5–15 tons/acre
- Sunn hemp: 4.5 tons/acre
- Switchgrass: 4–6 tons/acre

Genetically Modified Varieties

- Energy Sorghum

Biomass Conversion

Thermal Conversion

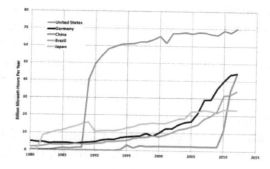

Trends in the top five countries generating electricity from biomass

Biomass briquettes are an example fuel for production of dendrothermal energy

Thermal conversion processes use heat as the dominant mechanism to convert biomass into another chemical form. The basic alternatives of combustion (torrefaction, pyrolysis, and gasification) are separated principally by the extent to which the chemical reactions involved are allowed to proceed (mainly controlled by the availability of oxygen and conversion temperature).

Energy created by burning biomass (fuel wood) is particularly suited for countries where the fuel wood grows more rapidly, e.g. tropical countries. There are a number of other less common, more experimental or proprietary thermal processes that may offer benefits such as hydrothermal upgrading (HTU) and hydroprocessing. Some have been developed for use on high moisture content biomass, including aqueous slurries, and allow them to be converted into more convenient forms. Some of the applications of thermal conversion are combined heat and power (CHP) and co-firing. In a typical dedicated biomass power plant, efficiencies range from 20–27% (higher heating value basis). Biomass cofiring with coal, by contrast, typically occurs at efficiencies near those of the coal combustor (30–40%, higher heating value basis).

Chemical Conversion

A range of chemical processes may be used to convert biomass into other forms, such as to produce a fuel that is more conveniently used, transported or stored, or to exploit some property of the process itself. Many of these processes are based in large part on similar coal-based processes, such as Fischer-Tropsch synthesis, methanol production, olefins (ethylene and propylene), and similar chemical or fuel feedstocks. In most cases, the first step involves gasification, which step generally is the most expensive and involves the greatest technical risk. Biomass is more difficult to feed into a pressure vessel than coal or any liquid. Therefore, biomass gasification is frequently done at atmospheric pressure and causes combustion of biomass to produce a combustible gas consisting of carbon monoxide, hydrogen, and traces of methane. This gas mixture, called a producer gas, can provide fuel for various vital processes, such as internal combustion engines, as well as substitute for furnace oil in direct heat applications. Because any biomass material can undergo gasification, this process is far more attractive than ethanol or biomass production, where only particular biomass materials can be used to produce a fuel. In addition, biomass gasification is a desirable process due to the ease at which it can convert solid waste (such as wastes available on a farm) into producer gas, which is a very usable fuel.

Conversion of biomass to biofuel can also be achieved via selective conversion of individual components of biomass. For example, cellulose can be converted to intermediate platform chemical such a sorbitol, glucose, hydroxymethylfurfural etc. These chemical are then further reacted to produce hydrogen or hydrocarbon fuels.

Biomass also has the potential to be converted to multiple commodity chemicals. Halomethanes have successfully been by produced using a combination of A. fermentans and engineered S. cerevisiae. This method converts NaX salts and unprocessed biomass such as switchgrass, sugarcane, corn stover, or poplar into halomethanes. S-adenosylmethionine which is naturally occurring in S. cerevisiae allows a methyl group to be transferred. Production levels of 150 mg $L^{-1}H^{-1}$ iodomethane were achieved. At these levels roughly 173000L of capacity would need to be operated just to replace the United States' need for iodomethane. However, an advantage of this method is that it uses NaI rather than I2; NaI is significantly less hazardous than I2. This method may be applied to produce ethylene in the future.

Other chemical processes such as converting straight and waste vegetable oils into biodiesel is transesterification.

Biochemical Conversion

As biomass is a natural material, many highly efficient biochemical processes have developed in nature to break down the molecules of which biomass is composed, and many of these biochemical conversion processes can be harnessed.

Biochemical conversion makes use of the enzymes of bacteria and other microorganisms to break down biomass. In most cases, microorganisms are used to perform the conversion process: anaerobic digestion, fermentation, and composting.

Electrochemical Conversion

In addition to combustion, bio-mass/bio-fuels can be directly converted to electrical energy via electrochemical oxidation of the material. This can be performed directly in a direct carbon fuel cell, direct ethanol fuel cell or a microbial fuel cell. The fuel can also be consumed indirectly via a fuel cell system containing a reformer which converts the bio-mass into a mixture of CO and H_2 before it is consumed in the fuel cell.

In the United States

The biomass power generating industry in the United States, which consists of approximately 11,000 MW of summer operating capacity actively supplying power to the grid, produces about 1.4 percent of the U.S. electricity supply.

Currently, the New Hope Power Partnership is the largest biomass power plant in the U.S. The 140 MW facility uses sugarcane fiber (bagasse) and recycled urban wood as fuel to generate enough power for its large milling and refining operations as well as to supply electricity for nearly 60,000 homes.

Second-generation Biofuels

Second-generation biofuels were not (in 2010) produced commercially, but a considerable number of research activities were taking place mainly in North America, Europe and also in some emerging countries. These tend to use feedstock produced by rapidly reproducing enzymes or bacteria from various sources including excrement grown in Cell cultures or hydroponics There is huge potential for second generation biofuels but non-edible feedstock resources are highly under-utilized.

Environmental Impact

Using biomass as a fuel produces air pollution in the form of carbon monoxide, carbon dioxide, NOx (nitrogen oxides), VOCs (volatile organic compounds), particulates and other pollutants at levels above those from traditional fuel sources such as coal or natural gas in some cases (such as with indoor heating and cooking). Utilization of wood biomass as a fuel can also produce fewer particulate and other pollutants than open burning as seen in wildfires or direct heat applications.

Black carbon – a pollutant created by combustion of fossil fuels, biofuels, and biomass – is possibly the second largest contributor to global warming. In 2009 a Swedish study of the giant brown haze that periodically covers large areas in South Asia determined that it had been principally produced by biomass burning, and to a lesser extent by fossil-fuel burning. Researchers measured a significant concentration of ^{14}C, which is associated with recent plant life rather than with fossil fuels.

Biomass power plant size is often driven by biomass availability in close proximity as transport costs of the (bulky) fuel play a key factor in the plant's economics. It has to be noted, however, that rail and especially shipping on waterways can reduce transport costs significantly, which has led to a global biomass market. To make small plants of 1 MW_{el} economically profitable those power plants need to be equipped with technology that is able to convert biomass to useful electricity with high efficiency such as ORC technology, a cycle similar to the water steam power process just with an organic working medium. Such small power plants can be found in Europe.

On combustion, the carbon from biomass is released into the atmosphere as carbon dioxide (CO_2). The amount of carbon stored in dry wood is approximately 50% by weight. However, according to the Food and Agriculture Organization of the United Nations, plant matter used as a fuel can be replaced by planting for new growth. When the biomass is from forests, the time to recapture the carbon stored is generally longer, and the carbon storage capacity of the forest may be reduced overall if destructive forestry techniques are employed.

Industry professionals claim that a range of issues can affect a plant's ability to comply with emissions standards. Some of these challenges, unique to biomass plants, include inconsistent fuel supplies and age. The type and amount of the fuel supply are completely reliant factors; the fuel can be in the form of building debris or agricultural waste (such as deforestation of invasive species or orchard trimmings). Furthermore, many of the biomass plants are old, use outdated technology and were not built to comply with today's stringent standards. In fact, many are based on technologies developed during the term of U.S. President Jimmy Carter, who created the United States Department of Energy in 1977.

The U.S. Energy Information Administration projected that by 2017, biomass is expected to be about twice as expensive as natural gas, slightly more expensive than nuclear power, and much less expensive than solar panels. In another EIA study released, concerning the government's plan to implement a 25% renewable energy standard by 2025, the agency assumed that 598 million tons of biomass would be available, accounting for 12% of the renewable energy in the plan.

The adoption of biomass-based energy plants has been a slow but steady process. Between the years of 2002 and 2012 the production of these plants has increased 14%. In the United States, alternative electricity-production sources on the whole generate about 13% of power; of this fraction, biomass contributes approximately 11% of the alternative production. According to a study conducted in early 2012, of the 107 operating biomass plants in the United States, 85 have been cited by federal or state regulators for the violation of clean air or water standards laws over the past 5 years. This data also includes minor infractions.

Despite harvesting, biomass crops may sequester carbon. For example, soil organic carbon has been observed to be greater in switchgrass stands than in cultivated cropland soil, especially at depths below 12 inches. The grass sequesters the carbon in its increased root biomass. Typically, perennial crops sequester much more carbon than annual crops due to much greater non-harvested living biomass, both living and dead, built up over years, and much less soil disruption in cultivation.

The proposal that biomass is carbon-neutral put forward in the early 1990s has been superseded by more recent science that recognizes that mature, intact forests sequester carbon more effectively than cut-over areas. When a tree's carbon is released into the atmosphere in a single pulse, it contributes to climate change much more than woodland timber rotting slowly over decades. Current studies indicate that "even after 50 years the forest has not recovered to its initial carbon storage" and "the optimal strategy is likely to be protection of the standing forest".

The pros and cons of biomass usage regarding carbon emissions may be quantified with the ILUC factor. There is controversy surrounding the usage of the ILUC factor.

Forest-based biomass has recently come under fire from a number of environmental organizations, including Greenpeace and the Natural Resources Defense Council, for the harmful impacts it can have on forests and the climate. Greenpeace recently released a report entitled "Fuelling a BioMess" which outlines their concerns around forest-based biomass. Because any part of the tree can be burned, the harvesting of trees for energy production encourages Whole-Tree Harvesting, which removes more nutrients and soil cover than regular harvesting, and can be harmful to the long-term health of the forest. In some jurisdictions, forest biomass removal is increasingly involving elements essential to functioning forest ecosystems, including standing trees, naturally disturbed forests and remains of traditional logging operations that were previously left in the forest. Environmental groups also cite recent scientific research which has found that it can take many decades for the carbon released by burning biomass to be recaptured by regrowing trees, and even longer in low productivity areas; furthermore, logging operations may disturb forest soils and cause them to release stored carbon. In light of the pressing need to reduce greenhouse gas emissions in the short term in order to mitigate the effects of climate change, a number of environmental groups are opposing the large-scale use of forest biomass in energy production.

Supply Chain Issues

With the seasonality of biomass supply and a great variability in sources, supply chains play a key role in cost-effective delivery of bioenergy. There are several potential challenges peculiar to bioenergy supply chains:

Technical issues

- Inefficiencies of conversion
- Storage methods for seasonal availability
- Complex multi-component constituents incompatible with maximizing efficiency of single purpose use
- High water content
- Conflicting decisions (technologies, locations, and routes)
- Complex location analysis (source points, inventory facilities, and production plants)

Logistic issues

- Seasonal availability leading to storage challenges and/or seasonally idle facilities
- Low bulk-density and/or high water content
- Finite productivity per area and/or time incompatible with conventional approach to economy of scale focusing on maximizing facility size

Financial issues

- The limits for the traditional approach to economy of scale which focuses on maximizing single facility size
- Unavailability and complexity of life cycle costing data
- Lack of required transport infrastructure
- Limited flexibility or inflexibility to energy demand
- Risks associated with new technologies (insurability, performance, rate of return)
- Extended market volatilities (conflicts with alternative markets for biomass)
- Difficult or impossible to use financial hedging methods to control cost

Social issues

- Lack of participatory decision making
- Lack of public/community awareness
- Local supply chain impacts vs. global benefits
- Health and safety risks
- Extra pressure on transport sector
- Decreasing the esthetics of rural areas

Policy and regulatory issues

- Impact of fossil fuel tax on biomass transport
- Lack of incentives to create competition among bioenergy producers
- Focus on technology options and less attention to selection of biomass materials
- Lack of support for sustainable supply chain solutions

Institutional and organizational issues

- Varied ownership arrangements and priorities among supply chain parties
- Lack of supply chain standards
- Impact of organizational norms and rules on decision making and supply chain coordination
- Immaturity of change management practices in biomass supply chains

Second Generation Biofuels

Second generation biofuels, also known as advanced biofuels, are fuels that can be manufactured from various types of biomass. Biomass is a wide-ranging term meaning any source of organic carbon that is renewed rapidly as part of the carbon cycle. Biomass is derived from plant materials but can also include animal materials.

First generation biofuels are made from the sugars and vegetable oils found in arable crops, which can be easily extracted using conventional technology. In comparison, second generation biofuels are made from lignocellulosic biomass or woody crops, agricultural residues or waste, which makes it harder to extract the required fuel.

Introduction

Second generation biofuel technologies have been developed because first generation biofuels manufacture has important limitations. First generation biofuel processes are useful but limited in most cases: there is a threshold above which they cannot produce enough biofuel without threatening food supplies and biodiversity. Many first generation biofuels depend on subsidies and are not cost competitive with existing fossil fuels such as oil, and some of them produce only limited greenhouse gas emissions savings. When taking emissions from production and transport into account, life-cycle assessment from first generation biofuels frequently approach those of traditional fossil fuels.

Second generation biofuels can help solve these problems and can supply a larger proportion of global fuel supply sustainably, affordably, and with greater environmental benefits.

First generation bioethanol is produced by fermenting plant-derived sugars to ethanol, using a similar process to that used in beer and wine-making. This requires the use of 'food' crops, such as sugar cane, corn, wheat, and sugar beet. These crops are required for food, so, if too much biofuel is made from them, food prices could rise and shortages might be experienced in some countries. Corn, wheat, and sugar beet can also require high agricultural inputs in the form of fertilizers, which limit the greenhouse gas reductions that can be achieved. Biodiesel produced by transesterification from rapeseed oil, palm oil, or other plant oils is also considered a first generation biofuel.

The goal of second generation biofuel processes is to extend the amount of biofuel that can be produced sustainably by using biomass consisting of the residual non-food parts of current crops, such as stems, leaves and husks that are left behind once the food crop has been extracted, as well as other crops that are not used for food purposes (non-food crops), such as switchgrass, grass, jatropha, whole crop maize, miscanthus and cereals that bear little grain, and also industry waste such as woodchips, skins and pulp from fruit pressing, etc.

The problem that second generation biofuel processes are addressing is to extract useful feedstocks from this woody or fibrous biomass, where the useful sugars are locked in by lignin, hemicellulose and cellulose. All plants contain lignin, hemicellulose and cellulose. These are complex carbohydrates (molecules based on sugar). Lignocellulosic ethanol is made by freeing the sugar molecules from cellulose using enzymes, steam heating, or other pre-treatments. These sugars can then be fermented to produce ethanol in the same way as first generation bioethanol production. The by-product of this process is lignin. Lignin can be burned as a carbon neutral fuel to produce heat and power for the processing plant and possibly for surrounding homes and businesses. Thermochemical processes (liquefaction) in hydrothermal media can produce liquid oily products from a wide range of feedstock that has a potential to replace or augment fuels. However, these liquid products fall short of diesel or biodiesel standards. Upgrading liquefaction products through one or many physical or chemical processes may improve properties for use as fuel.

Second Generation Technology

The following subsections describe the main second generation routes currently under development.

Thermochemical Routes

Carbon-based materials can be heated at high temperatures in the absence (pyrolysis) or presence of oxygen, air and/or steam (gasification).

These thermochemical processes both yield a combustible gas and a solid char. The gas can be fermented or chemically synthesised into a range of fuels, including ethanol, synthetic diesel or jet fuel.

There are also lower temperature processes in the region of 150-374 °C, that produce sugars by decomposing the biomass in water with or without additives.

Gasification

Gasification technologies are well established for conventional feedstocks such as coal and crude oil. Second generation gasification technologies include gasification of forest and agricultural residues, waste wood, energy crops and black liquor. Output is normally syngas for further synthesis to e.g. Fischer-Tropsch products including diesel fuel, biomethanol, BioDME (dimethyl ether), gasoline via catalytic conversion of dimethyl ether, or biomethane (synthetic natural gas). Syngas can also be used in heat production and for generation of mechanical and electrical power via gas motors or gas turbines.

Pyrolysis

Pyrolysis is a well established technique for decomposition of organic material at elevated temperatures in the absence of oxygen. In second generation biofuels applications forest and agricultural residues, wood waste and energy crops can be used as feedstock to produce e.g. bio-oil for fuel oil applications. Bio-oil typically requires significant additional treatment to render it suitable as a refinery feedstock to replace crude oil.

Torrefaction

Torrefaction is a form of pyrolysis at temperatures typically ranging between 200-320 °C. Feedstocks and output are the same as for pyrolysis.

Biochemical Routes

Chemical and biological processes that are currently used in other applications are being adapted for second generation biofuels. Biochemical processes typically employ pre-treatment to accelerate the hydrolysis process, which separates out the lignin, hemicellulose and cellulose. Once these ingredients are separated, the cellulose fractions can be fermented into alcohols.

Feedstocks are energy crops, agricultural and forest residues, food industry and municipal bio-waste and other biomass containing sugars. Products include alcohols (such as ethanol and butanol) and other hydrocarbons for transportation use.

Types of Biofuel

The following second generation biofuels are under development, although most or all of these biofuels are synthesized from intermediary products such as syngas using methods that are identical in processes involving conventional feedstocks, first generation and second generation biofuels. The distinguishing feature is the technology involved in producing the intermediary product, rather than the ultimate off-take.

A process producing liquid fuels from gas (normally syngas) is called a Gas-to-Liquid (GtL) process. When biomass is the source of the gas production the process is also referred to as Biomass-To-Liquids (BTL).

From Syngas Using Catalysis

- Biomethanol can be used in methanol motors or blended with petrol up to 10-20% without any infrastructure changes.

- BioDME can be produced from Biomethanol using catalytic dehydration or it can be produced directly from syngas using direct DME synthesis. DME can be used in the compression ignition engine.

- Bio-derived gasoline can be produced from DME via high-pressure catalytic condensation reaction. Bio-derived gasoline is chemically indistinguishable from petroleum-derived gasoline and thus can be blended into the U.S. gasoline pool.

- Biohydrogen can be used in fuel cells to produce electricity.

- Mixed Alcohols (i.e., mixture of mostly ethanol, propanol, and butanol, with some pentanol, hexanol, heptanol, and octanol). Mixed alcohols are produced from syngas with several classes of catalysts. Some have employed catalysts similar to those used for methanol. Molybdenum sulfide catalysts were discovered at Dow Chemical and have received considerable attention. Addition of cobalt sulfide to the catalyst formulation was shown to enhance performance. Molybdenum sulfide catalysts have been well studied but have yet to find widespread use. These catalysts have been a focus of efforts at the U.S. Department of Energy's Biomass Program in the Thermochemical Platform. Noble metal catalysts have also been shown to produce mixed alcohols. Most R&D in this area is concentrated in producing mostly ethanol. However, some fuels are marketed as mixed alcohols Mixed alcohols are superior to pure methanol or ethanol, in that the higher alcohols have higher energy content. Also, when blending, the higher alcohols increase compatibility of gasoline and ethanol, which increases water tolerance and decreases evaporative emissions. In addition, higher alcohols have also lower heat of vaporization than ethanol, which is important for cold starts.

- Biomethane (or Bio-SNG) via the Sabatier reaction

From Syngas Using Fischer-Tropsch

The Fischer-Tropsch (FT) process is a Gas-to-Liquid (GtL) process. When biomass is the source of the gas production the process is also referred to as Biomass-To-Liquids (BTL). A disadvantage of this process is the high energy investment for the FT synthesis and consequently, the process is not yet economic.

FT diesel can be mixed with fossil diesel at any percentage without need for infrastructure change and moreover, synthetic kerosene can be produced

Biocatalysis

- Biohydrogen might be accomplished with some organisms that produce hydrogen directly under certain conditions. Biohydrogen can be used in fuel cells to produce electricity.

- Butanol and Isobutanol via recombinant pathways expressed in hosts such as E. coli and yeast, butanol and isobutanol may be significant products of fermentation using glucose as a carbon and energy source.

- DMF (2,5-Dimethylfuran). Recent advances in producing DMF from fructose and glucose using catalytic biomass-to-liquid process have increased its attractiveness.

Other Processes

- HTU (Hydro Thermal Upgrading) diesel is produced from wet biomass. It can be mixed with fossil diesel in any percentage without need for infrastructure.

- Wood diesel. A new biofuel was developed by the University of Georgia from woodchips. The oil is extracted and then added to unmodified diesel engines. Either new plants are used or planted to replace the old plants. The charcoal byproduct is put back into the soil as a fertilizer. According to the director Tom Adams since carbon is put back into the soil, this biofuel can actually be carbon negative not just carbon neutral. Carbon negative decreases carbon dioxide in the air reversing the greenhouse effect not just reducing it.

Feedstocks

Second generation biofuel feedstocks include cereal and sugar crops, specifically grown energy crops, agricultural and municipal wastes, cultivated and waste oils, and algae. Land use, existing biomass industries and relevant conversion technologies must be considered when evaluating suitability of developing biomass as feedstock for energy.

Energy Crops

Plants are made from lignin, hemicellulose and cellulose; second generation technology uses one, two or all of these components. Common lignocellulosic energy crops include wheat straw, Miscanthus, short rotation coppice poplar and willow. However, each offers different opportunities and no one crop can be considered 'best' or 'worst'.

Municipal Solid Waste

Municipal Solid Waste comprises a very large range of materials, and total waste arisings are increasing. In the UK, recycling initiatives decrease the proportion of waste going straight for disposal, and the level of recycling is increasing each year. However, there remains significant opportunities to convert this waste to fuel via gasification or pyrolysis.

Green Waste

Green waste such as forest residues or garden or park waste may be used to produce biofuel via different routes. Examples include Biogas captured from biodegradable green waste, and gasification or hydrolysis to syngas for further processing to biofuels via catalytic processes.

Black Liquor

Black liquor, the spent cooking liquor from the kraft process that contains concentrated lignin and hemicellulose, may be gasified with very high conversion efficiency and greenhouse gas reduction potential to produce syngas for further synthesis to e.g. biomethanol or BioDME.

Greenhouse Gas Emissions

Lignocellulosic biofuels reduces greenhouse gas emissions with 60-90% when compared with fossil petroleum (Börjesson.P. et al. 2013. Dagens och framtidens hållbara biodrivmedel), which is on par with the better of current biofuels of the first generation, where typical best values currently is 60-80%. In 2010, average savings of biofuels used within EU was 60% (Hamelinck.C. et al. 2013 Renewable energy progress and biofuels sustainability, Report for the European Commission). In 2013, 70% of the biofuels used in Sweden reduced emissions with 66% or higher. (Energimyndigheten 2014. Hållbara biodrivmedel och flytande biobränslen 2013).

Commercial Development

An operating lignocellulosic ethanol production plant is located in Canada, run by Iogen Corporation. The demonstration-scale plant produces around 700,000 litres of bioethanol each year. A commercial plant is under construction. Many further lignocellulosic ethanol plants have been proposed in North America and around the world.

The Swedish specialty cellulose mill Domsjö Fabriker in Örnsköldsvik, Sweden develops a biorefinery using Chemrec's black liquor gasification technology. When commissioned in 2015 the biorefinery will produce 140,000 tons of biomethanol or 100,000 tons of BioDME per year, replacing

2% of Sweden's imports of diesel fuel for transportation purposes. In May 2012 it was revealed that Domsjö pulled out of the project, effectively killing the effort.

In the UK, companies like INEOS Bio and British Airways are developing advanced biofuel refineries, which are due to be built by 2013 and 2014 respectively. Under favourable economic conditions and strong improvements in policy support, NNFCC projections suggest advanced biofuels could meet up to 4.3 per cent of the UK's transport fuel by 2020 and save 3.2 million tonnes of CO_2 each year, equivalent to taking nearly a million cars off the road.

Helsinki, Finland, 1 February 2012 - UPM is to invest in a biorefinery producing biofuels from crude tall oil in Lappeenranta, Finland. The industrial scale investment is the first of its kind globally. The biorefinery will produce annually approximately 100,000 tonnes of advanced second generation biodiesel for transport. Construction of the biorefinery will begin in the summer of 2012 at UPM's Kaukas mill site and be completed in 2014. UPM's total investment will amount to approximately EUR 150 million.

Calgary, Alberta, 30 April 2012 – Iogen Energy Corporation has agreed to a new plan with its joint owners Royal Dutch Shell and Iogen Corporation to refocus its strategy and activities. Shell continues to explore multiple pathways to find a commercial solution for the production of advanced biofuels on an industrial scale, but the company will NOT pursue the project it has had under development to build a larger scale cellulosic ethanol facility in southern Manitoba.

"Drop-in" Biofuels

So-called "drop-in" biofuels can be defined as "liquid bio-hydrocarbons that are functionally equivalent to petroleum fuels and are fully compatible with existing petroleum infrastructure".

There is considerable interest in developing advanced biofuels that can be readily integrated in the existing petroleum fuel infrastructure - i.e. "dropped-in" - particularly by sectors such as aviation, where there are no real alternatives to sustainably produced biofuels for low carbon emitting fuel sources. Drop-in biofuels by definition should be fully fungible and compatible with the large existing "petroleum-based" infrastructure.

According to a recent report published by the IEA Bioenergy Task 39, entitled "The potential and challenges of drop-in biofuels", there are several ways to produce drop-in biofuels that are functionally equivalent to petroleum-derived transportation fuel blendstocks. These are discussed within three major sections of the full report and include:

- oleochemical processes, such as the hydroprocessing of lipid feedstocks obtained from oilseed crops, algae or tallow;
- thermochemical processes, such as the thermochemical conversion of biomass to fluid intermediates (gas or oil) followed by catalytic upgrading and hydroprocessing to hydrocarbon fuels; and
- biochemical processes, such as the biological conversion of biomass (sugars, starches or lignocellulose-derived feedstocks) to longer chain alcohols and hydrocarbons.

A fourth category is also briefly described that includes "hybrid" thermochemical/biochemical technologies such as fermentation of synthesis gas and catalytic reforming of sugars/carbohydrates.

The report concludes by stating:

"Tremendous entrepreneurial activity to develop and commercialize drop-in biofuels from aquatic and terrestrial feedstocks has taken place over the past several years. However, despite these efforts, drop-in biofuels represent only a small percentage (around 2%) of global biofuel markets. (...) Due to the increased processing and resource requirements (e.g., hydrogen and catalysts) needed to make drop-in biofuels as compared to conventional biofuels, large scale production of cost-competitive drop-in biofuels is not expected to occur in the near to midterm. Rather, dedicated policies to promote development and commercialization of these fuels will be needed before they become significant contributors to global biofuels production. Currently, no policies (e.g., tax breaks, subsidies etc.) differentiate new, more fungible and infrastructure ready drop-in type biofuels from less infrastructure compatible oxygenated biofuels. (...) Thus, while tremendous technical progress has been made in developing and improving the various routes to drop-in fuels, supportive policies directed specifically towards the further development of drop-in biofuels are likely to be needed to ensure their future commercial success".

Biomass Briquettes

Biomass briquettes are a biofuel substitute to coal and charcoal. Briquettes are mostly used in the developing world, where cooking fuels are not as easily available. There has been a move to the use of briquettes in the developed world, where they are used to heat industrial boilers in order to produce electricity from steam. The briquettes are cofired with coal in order to create the heat supplied to the boiler.

Briquette made by a Ruf briquetter out of hay

Straw or hay briquettes

Ogatan, Japanese charcoal briquettes made from sawdust briquettes *(Ogalite)*.

Quick Grill Briquette made from coconut shell

Composition and Production

Biomass briquettes, mostly made of green waste and other organic materials, are commonly used for electricity generation, heat, and cooking fuel. These compressed compounds contain various organic materials, including rice husk, bagasse, ground nut shells, municipal solid waste, agricultural waste. The composition of the briquettes varies by area due to the availability of raw materials. The raw materials are gathered and compressed into briquette in order to burn longer and make transportation of the goods easier. These briquettes are very different from charcoal because they do not have large concentrations of carbonaceous substances and added materials. Compared to fossil fuels, the briquettes produce low net total greenhouse gas emissions because the materials used are already a part of the carbon cycle.

One of the most common variables of the biomass briquette production process is the way the biomass is dried out. Manufacturers can use torrefaction, carbonization, or varying degrees of pyrolysis. Researchers concluded that torrefaction and carbonization are the most efficient forms of drying out biomass, but the use of the briquette determines which method should be used.

Compaction is another factor affecting production. Some materials burn more efficiently if compacted at low pressures, such as corn stover grind. Other materials such as wheat and barley-straw require high amounts of pressure to produce heat. There are also different press technologies that can be used. A piston press is used to create solid briquettes for a wide array of purposes. Screw extrusion is used to compact biomass into loose, homogeneous briquettes that are substituted for coal in cofiring. This technology creates a toroidal, or doughnut-like, briquette. The hole in the center of the briquette allows for a larger surface area, creating a higher combustion rate.

History

People have been using biomass briquettes in Nepal since before recorded history. Though inefficient, the burning of loose biomass created enough heat for cooking purposes and keeping warm. The first commercial production plant was created in 1982 and produced almost 900 metric tons of biomass. In 1984, factories were constructed that incorporated vast improvements on efficiency and the quality of briquettes. They used a combination of rice husks and molasses. The King Mahendra Trust for Nature Conservation (KMTNC) along with the Institute for Himalayan Conservation (IHC) created a mixture of coal and biomass in 2000 using a unique rolling machine.

Japanese Ogalite

In 1925, Japan independently started developing technology to harness the energy from sawdust briquettes, known as *"Ogalite"*. Between 1964 and 1969, Japan increased production fourfold by incorporating screw press and piston press technology. The member enterprise of 830 or more existed in the 1960s. The new compaction techniques incorporated in these machines made briquettes of higher quality than those in Europe. As a result, European countries bought the licensing agreements and now manufacture Japanese designed machines.

Cofiring

Cofiring relates to the combustion of two different types of materials. The process is primarily used to decrease CO_2 emissions despite the resulting lower energy efficiency and higher variable cost. The combination of materials usually contains a high carbon emitting substance such as coal and a lesser CO_2 emitting material such as biomass. Even though CO_2 will still be emitted through the combustion of biomass, the net carbon emitted is nearly negligible. This is due to the fact that the material gathered for the composition of the briquettes are still contained in the carbon cycle whereas fossil fuel combustion releases CO_2 that has been sequestered for millennia. Boilers in power plants are traditionally heated by the combustion of coal, but if co-firing were to be implemented, then the CO_2 emissions would decrease while still maintaining the heat inputted to the boiler. Implementing cofiring would require few modifications to the current characteristics to power plants, as only the fuel for the boiler would be altered. A moderate investment would be required for implementing biomass briquettes into the combustion process.

Cofiring is considered the most cost-efficient means of biomass. A higher combustion rate will occur when cofiring is implemented in a boiler when compared to burning only biomass. The compressed biomass is also much easier to transport since it is more dense, therefore allowing more biomass to be transported per shipment when compared to loose biomass. Some sources agree that a near-term solution for the greenhouse gas emission problem may lie in cofiring.

Compared to Coal

The use of biomass briquettes has been steadily increasing as industries realize the benefits of decreasing pollution through the use of biomass briquettes. Briquettes provide higher calorific value per dollar than coal when used for firing industrial boilers. Along with higher calorific value, biomass briquettes on average saved 30–40% of boiler fuel cost. But other sources suggest that co-firing is more expensive due to the widespread availability of coal and its low cost. However, in the long run, briquettes can only limit the use of coal to a small extent, but it is increasingly being pursued by industries and factories all over the world. Both raw materials can be produced or mined domestically in the United States, creating a fuel source that is free from foreign dependence and less polluting than raw fossil fuel incineration.

Environmentally, the use of biomass briquettes produces much fewer greenhouse gases, specifically, 13.8% to 41.7% CO_2 and NO_x. There was also a reduction from 11.1% to 38.5% in SO2 emissions when compared to coal from three different leading producers, EKCC Coal, Decanter Coal, and Alden Coal. Biomass briquettes are also fairly resistant to water degradation, an improvement over the difficulties encountered with the burning of wet coal. However, the briquettes are best used only as a supplement to coal. The use of cofiring creates an energy that is not as high as pure coal, but emits fewer pollutants and cuts down on the release of previously sequestered carbon. The continuous release of carbon and other greenhouse gasses into the atmosphere leads to an increase in global temperatures. The use of cofiring does not stop this process but decreases the relative emissions of coal power plants.

Use in Developing World

The Legacy Foundation has developed a set of techniques to produce biomass briquettes through artisanal production in rural villages that can be used for heating and cooking. These techniques were recently pioneered by Virunga National Park in eastern Democratic Republic of Congo, following the massive destruction of the Mountain Gorilla habitat for charcoal.

Pangani, Tanzania, is an area covered in coconut groves. After harvesting the meat of the coconut, the indigenous people would litter the ground with the husks, believing them to be useless. The husks later became a profit center after it was discovered that coconut husks are well suited to be the main ingredient in bio briquettes. This alternative fuel mixture burns incredibly efficiently and leaves little residue, making it a reliable source for cooking in the undeveloped country. The developing world has always relied on the burning biomass due it its low cost and availability anywhere there is organic material. The briquette production only improves upon the ancient practice by increasing the efficiency of pyrolysis.

Two major components of the developing world are China and India. The economies are rapidly increasing due to cheap ways of harnessing electricity and emitting large amounts of carbon dioxide. The Kyoto Protocol attempted to regulate the emissions of the three different worlds, but there were disagreements as to which country should be penalized for emissions based on its previous and future emissions. The United States has been the largest emitter but China has recently become the largest per capita. The United States had emitted a rigorous amount of carbon dioxide during its development and the developing nations argue that they should not be forced to meet the requirements. At the lower end, the undeveloped nations believe that they have little responsibility for what has been done to the carbon dioxide levels. The major use of biomass briquettes in India, is in industrial applications usually to produce steam. A lot of conversions of boilers from FO to biomass briquettes have happened over the past decade. A vast majority of those projects are registered under CDM (Kyoto Protocol), which allows for users to get carbon credits.

The use of biomass briquettes is strongly encouraged by issuing carbon credits. One carbon credit is equal to one free ton of carbon dioxide to be emitted into the atmosphere. India has started to replace charcoal with biomass briquettes in regards to boiler fuel, especially in the southern parts of the country because the biomass briquettes can be created domestically, depending on the availability of land. Therefore, constantly rising fuel prices will be less influential in an economy if sources of fuel can be easily produced domestically. Lehra Fuel Tech Pvt Ltd is approved by Indian Renewable Energy Development Agency (IREDA), is one of the largest briquetting machine manufacturers from Ludhiana, India.

In the African Great Lakes region, work on biomass briquette production has been spearheaded by a number of NGOs with GVEP (Global Village Energy Partnership) taking a lead in promoting briquette products and briquette entrepreneurs in the three Great Lakes countries; namely, Kenya, Uganda and Tanzania. This has been achieved by a five-year EU and Dutch government sponsored project called DEEP EA (Developing Energy Enterprises Project East Africa) . The main feed stock for briquettes in the East African region has mainly been charcoal dust although alternative like sawdust, bagasse, coffee husks and rice husks have also been used.

Use in Developed World

Coal is the largest carbon dioxide emitter per unit area when it comes to electricity generation. It is also the most common ingredient in charcoal. There has been a recent push to replace the burning of fossil fuels with biomass. The replacement of this nonrenewable resource with biological waste would lower the carbon footprint of grill owners and lower the overall pollution of the world. Citizens are also starting to manufacture briquettes at home. The first machines would create briquettes for homeowners out of compressed sawdust, however, current machines allow for briquette production out of any sort of dried biomass.

Arizona has also taken initiative to turn waste biomass into a source of energy. Waste cotton and pecan material used to provide a nesting ground for bugs that would destroy the new crops in the spring. To stop this problem farmers buried the biomass, which quickly led to soil degradation. These materials were discovered to be a very efficient source of energy and took care of issues that had plagued farms.

The United States Department of Energy has financed several projects to test the viability of biomass briquettes on a national scale. The scope of the projects is to increase the efficiency of gasifiers as well as produce plans for production facilities.

Criticism

Biomass is composed of organic materials, therefore, large amounts of land are required to produce the fuel. Critics argue that the use of this land should be utilized for food distribution rather than crop degradation. Also, climate changes may cause a harsh season, where the material extracted will need to be swapped for food rather than energy. The assumption is that the production of biomass decreases the food supply, causing an increase in world hunger by extracting the organic materials such as corn and soybeans for fuel rather than food.

The cost of implementing a new technology such as biomass into the current infrastructure is also high. The fixed costs with the production of biomass briquettes are high due to the new undeveloped technologies that revolve around the extraction, production and storage of the biomass. Technologies regarding extraction of oil and coal have been developing for decades, becoming more efficient with each year. A new undeveloped technology regarding fuel utilization that has no infrastructure built around makes it nearly impossible to compete in the current market.

References

- Mani, Sokhansanj, and L.G. Tabil. "Evaluation of compaction equations applied to four biomass species." University of Saskatchewan College of Engineering. Web. 30 Nov. 2010.

- "Biomass Briquetting: Technology and Practices - Introduction." Centre for Ecological Sciences INDIAN INSTITUTE OF SCIENCE BANGALORE. Web. 04 Dec. 2010.

- Yugo Isobe, Kimiko Yamada, Qingyue Wang, Kazuhiko Sakamoto, Iwao Uchiyama, Tsuguo Mizoguchi and Yanrong Zhou. "Measurement of Indoor Sulfur Dioxide Emission from Coal–Biomass Briquettes." springerlink.com. Web. 30 November 2010.

- Montross, Neathery, O'Daniel, Patil, Sowder and Darrell Taulbee. (2010). "Combustion of Briquettes and Fuels Pellets Prepared from Blends of Biomass and Fine Coal". International Coal Preparation 2010 Conference Proceeding (161-170). Google Books. Web. 29 November 2010

- "DOE Selects Projects to Advance Technologies for the Co-Production of Power and Hydrogen, Fuels or Chemicals from Coal-Biomass Feedstocks." United States Department of Energy. 18 Aug. 2010. Web. 04 Dec. 2010.

- Knight, Chris (2013). "Chapter 6 – Application of Microbial Fuel Cells to Power Sensor Networks for Ecological Monitoring". Wireless Sensor Networks and Ecological Monitoring. Smart Sensors, Measurement and Instrumentation. 3. pp. 151–178. doi:10.1007/978-3-642-36365-8_6. ISBN 978-3-642-36364-1.

- Starke, Linda (2009). State Of The World 2009: Into a Warming World: a WorldWatch Institute Report on Progress Toward a Sustainable Society. WW Norton & Company. ISBN 978-0-393-33418-0.

- "Learning About Renewable Energy". NREL's vision is to develop technology. National Renewable Energy Laboratory. Retrieved 4 April 2013.

- "Biomass for Electricity Generation". capacity of about 6.7 gigawatts in 2000 to about 10.4 gigawatts by 2020. U.S. Energy Information Administration (EIA). Retrieved 6 April 2013.

- Forest volume-to-biomass models and estimates of mass for live and standing dead trees of U.S. forests. (PDF). Retrieved on 2012-02-28.

- U.S. Energy Information Administration (April 2010). Annual Energy Outlook 2010 (PDF) (report no. DOE/EIA-0383(2010)). Washington, DC. National Energy Information Center. Retrieved September 27, 2012.

- Scheck, Justin; et al. (July 23, 2012). "Wood-Fired Plants Generate Violations". Wall Street Journal. Retrieved September 27, 2012.

- NRDC fact sheet lays out biomass basics, campaign calls for action to tell EPA our forests aren't fuel | Sasha Lyutse's Blog | Switchboard, from NRDC. Switchboard.nrdc.org (2011-05-02). Retrieved on 2012-02-28.

- Prasad, Ram. "SUSTAINABLE FOREST MANAGEMENT FOR DRY FORESTS OF SOUTH ASIA". Food and Agriculture Organization of the United Nations. Retrieved 11 August 2010.

- "Treetrouble: Testimonies on the Negative Impact of Large-scale Tree Plantations prepared for the sixth Conference of the Parties of the Framework Convention on Climate Change". Friends of the Earth International. Retrieved 11 August 2010.

- "Fuel Ethanol Production: GSP Systems Biology Research". U.S. Department of Energy Office of Science. April 19, 2010. Archived from the original on 2010-10-28. Retrieved 2010-08-02.

- "European Commission - PRESS RELEASES - Press release - State aid: Commission approves Swedish €55 million aid for "Domsjö" R&D project". Retrieved 22 September 2015.

- National Non-Food Crops Centre. "Pathways to UK Biofuels: A Guide to Existing and Future Options for Transport, NNFCC 10-035", Retrieved on 2011-06-27

- National Non-Food Crops Centre. "Advanced Biofuels: The Potential for a UK Industry, NNFCC 11-011", Retrieved on 2011-11-17

- Evans, G. "International Biofuels Strategy Project. Liquid Transport Biofuels - Technology Status Report, NNFCC 08-017", National Non-Food Crops Centre, 2008-04-14. Retrieved on 2011-02-16.

- National Non-Food Crops Centre. "Review of Technologies for Gasification of Biomass and Wastes, NNFCC 09-008", Retrieved on 2011-06-24

- National Non-Food Crops Centre. "Evaluation of Opportunities for Converting Indigenous UK Wastes to Fuels and Energy (Report), NNFCC 09-012", Retrieved on 2011-06-27

Use of Bioethanol in Sustainable Transport

This chapter serves as a source to understand sustainable transport and the use of bioethanol. Bio-Ethanol for Sustainable Transport (BEST) was a project that was introduced to support the usage of bioethanol as a vehicle fuel. The purpose of the use of bioethanol in transport is to have more flexible fuel vehicles and to decrease the amount of pollution caused by vehicles.

BioEthanol for Sustainable Transport (BEST) was a four-year project financially supported by the European Union for promoting the introduction and market penetration of bioethanol as a vehicle fuel, and the introduction and wider use of flexible-fuel vehicles and ethanol-powered vehicles on the world market. The project began in January 2006 and continued until the end of 2009, and had nine participating regions or cities in Europe, Brazil, and China.

Goals

The BEST project targets included the introduction of more than 10,000 flex-fuel or ethanol cars and 160 ethanol buses; to promote the opening of 135 E85 and 13 ED95 public fuel stations; and to promote the development and testing of hydrous E15 and anhydrous low ethanol blends with gasoline and diesel.

Participants

There were ten participating cities and regions, and several commercial partners. Stockholm (Sweden) was the coordinating city, and other participants were Basque Country and Madrid (Spain), the Biofuel Region in Sweden, Brandenburg (Germany), La Spezia (Italy), Nanyang (China), Rotterdam (Netherlands), São Paulo (Brazil), and Somerset (UK). The commercial partners were Ford Europe, Saab Automobile and several bioethanol suppliers.

Implemented Projects

Flexible-fuel Vehicles

A major activity in BEST was the promotion of E85 flexifuel vehicles (FFVs). During the project nine BEST sites introduced over 77,000 FFVs, far exceeding the original project's target of 10,000 vehicles. In 2008, out of the 170,000 flexifuel vehicles in operation in Europe, 45% of the vehicles operated at BEST sites; and out of 2,200 E85 pumps installed in the EU, 80% are found in the BEST countries. Sweden stands out with 70% of all flexifuel vehicles operating in the EU. BEST sites also evaluated both dedicated E85 pumps and flexifuel pumps and found very few problems.

Ethanol-powered Buses

The project included demonstration of two types of bioethanol-powered buses, a diesel engine Scania bus running on ED95 (sugarcane ethanol plus an ignition improver) and a Dongfeng bus capable of running on both E100 and petrol (flexible-fuel bus). Fuel pumps were also installed at bus depots in the five participating cities.

ED95-powered Scania OmniCity bus in Stockholm.

BEST demonstrated more than 138 bioethanol ED95 buses and 12 ED95 pumps at five sites, three in Europe, one in China and one in Brazil. These trials helped increase knowledge about bioethanol buses in the participating cities. An innovation within BEST was the demonstration of two dual-tank E100 buses developed by the Chinese vehicle producer Dongfeng Motor. All BEST sites will continue to drive their bioethanol buses in regular traffic and some cities are already planning to expand their fleets.

The trial demonstrations showed that ethanol-powered ED95 buses:

- reduce greenhouse gas emissions and local air pollution.

- are reliable and appreciated by drivers and passengers.

- cost more to purchase and operate than diesel buses.

- require more scheduled maintenance than diesel buses.

- Taxing fuel by volume instead of energy content penalises bioethanol buses.

- ED95 can be safely handled at depots and has potential for wider use in heavy vehicles such as trucks.

Brazil

Under the auspices of the BEST project, the first ED95 bus began operations in São Paulo city on December 2007 as a trial project. The bus is a Scania with a modified diesel engine capable of running with 95% hydrous ethanol with 5% ignition improver. Scania adjusted the compression ratio from 18:1 to 28:1, added larger fuel injection nozzles, and altered the injection timing.

President Luiz Inácio Lula da Silva and King Carl XVI Gustaf of Sweden inspecting one of the 400 buses running on ED95 in Stockholm

During the first year trial period performance and emissions were monitored by the National Reference Center on Biomass (CENBIO - Portuguese: *Centro Nacional de Referência em Biomassa*) at the Universidade de São Paulo, and compared with similar diesel models, with special attention to carbon monoxide and particulate matter emissions. Performance is also important as previous tests have shown a reduction in fuel economy of around 60% when E95 is compared to regular diesel.

In November 2009, a second ED95 bus began operating in São Paulo city. The bus was a Swedish Scania engine and chassis with a CAIO bus body. The second bus was scheduled to operate between Lapa and Vila Mariana, passing through Avenida Paulista, one of the main business centers of São Paulo city.

CENBIO laboratory tests found that as compared to diesel, carbon dioxide emissions are 80% lower with ED95, particulate drops by 90%, nitrogen oxide emissions are 62% lower, and there are no sulphur emissions. During the trial was observed that the first bus began presenting sudden halts of the engine in slow running. The problem manifested more frequently in hot days, when the ambient temperature reached 26 °C or more and on top of long grades. After analyzing carefully the problem in the engine's fuel power line, it was discovered that the bus was developed for the European temperate climate, where average temperatures are lower than in tropical climate. In hotter days, the temperature of the fuel line reached up to 58 °C, a temperature that could increase even more when the engine would be slow running. The excessive heating was causing the vaporization of the fuel in the power line of the engine. The solution found was to deviate the fuel return from the engine straight to the tank, and thus, adapting the engine to Brazilian climate conditions.

Based on the satisfactory results obtained during the 3-year trial operation of the two buses, in November 2010 the municipal government of São Paulo city signed an agreement with UNICA, Cosan, Scania and *Viação Metropolitana*", a local bus operator, to introduce a fleet of 50 ethanol-powered

ED95 buses by May 2011. The city's government objective is to reduce the carbon footprint of the city's bus fleet which is made of 15,000 diesel-powered buses, and the final goal is for the entire bus fleet to use only renewable fuels by 2018 . The first ethanol-powered buses were delivered in May 2011, and the 50 ethanol-powered ED95 buses will begin regular service in June 2011.

China

In Nanyang, Henan, a new type of bioethanol flexible-fuel bus capable of running on petrol or neat ethanol fuel (E100) was developed by Dongfeng Motor. The buses look like conventional buses and have two fuel tanks, one for petrol and one for E100. Two buses were demonstrated by local bioethanol producer Tianguan, who also supplied E100 for the buses. One fuel pump was set up for the trial. One of the buses uses a modified petrol-engine and the other uses a modified natural gas engine. The new bus types were developed to overcome import duties and are a low-cost alternative for Chinese cities seeking to introduce bioethanol to their public transport systems. Each E100 bus developed by Dongfeng costs around €35,000, which is €1,000 more expensive than a conventional petrol bus.

Italy

Three ED95 buses and one fuel pump was installed in La Spezia.

Spain

Five ED95 buses operated in Madrid and one fuel pump was installed.

Sweden

In Stockholm a total of 127 ED95 buses and five ED95 ethanol fuel stations were funded within the BEST project.

References

- BEST Cities and Regions (2009). "BioEthanol for Sustainable Transport: Results and recommendations from the European BEST project" (PDF). Environmental and Health Administration, City of Stockholm. Retrieved 2011-05-28.

- Fernando Saker (2009-11-13). "São Paulo recebe segundo ônibus movido a etanol" (in Portuguese). Centro Nacional de Referência em Biomassa. Retrieved 2011-05-28.

- Martha San Juan França (2011-05-26). "São Paulo ganha frota de ônibus a etanol". Brasil Econômico (in Portuguese). Retrieved 2011-05-27.

- Moreira, José Roberto & Sandra Maria Apolinario, Suani Teixeira Coelho (2010). "Ethanol Usage in Urban Public Transportation - Presentation of Results" (PDF). Brazilian Reference Center on Biomass (CENBIO). Retrieved 2011-05-29.

- "Substituição de diesel por etanol: vantagem que novos ônibus trazem à cidade de São Paulo". UNICA (in Portuguese). 2011-05-26. Retrieved 2011-05-27.

- Reuters (2010-11-25). "São Paulo terá primeira frota de ônibus movida a etanol". Terra Networks (in Portuguese). Retrieved 2010-11-27.

- Eduardo Magossi (2010-11-25). "Cosan fornecerá etanol para 50 ônibus de SP". O Estado de S. Paulo (in Portuguese). Retrieved 2010-11-27.

6

Various Issues Related to Biofuels

This chapter provides a thorough understanding of the distinctive issues relevant to biofuels. Biofuels have several social and environmental concerns; some of the major challenges faced are the food vs. fuel issue, sustainable biofuel and deforestation. The content within this chapter attempts to summarize the disadvantages that should be considered in the use of biofuels, in order to encourage responsible consumption of the same.

Issues Relating to Biofuels

There are various social, economic, environmental and technical issues with biofuel production and use, which have been discussed in the popular media and scientific journals. These include: the effect of moderating oil prices, the "food vs fuel" debate, poverty reduction potential, carbon emissions levels, sustainable biofuel production, deforestation and soil erosion, loss of biodiversity, effect on water resources, the possible modifications necessary to run the engine on biofuel, as well as energy balance and efficiency. The International Resource Panel, which provides independent scientific assessments and expert advice on a variety of resource-related themes, assessed the issues relating to biofuel use in its first report *Towards sustainable production and use of resources: Assessing Biofuels*. In it, it outlined the wider and interrelated factors that need to be considered when deciding on the relative merits of pursuing one biofuel over another. It concluded that not all biofuels perform equally in terms of their effect on climate, energy security and ecosystems, and suggested that environmental and social effects need to be assessed throughout the entire life-cycle.

Social and Economic Effects

Oil Price Moderation

The International Energy Agency's *World Energy Outlook 2006* concludes that rising oil demand, if left unchecked, would accentuate the consuming countries' vulnerability to a severe supply disruption and resulting price shock. The report suggested that biofuels may one day offer a viable alternative, but also that "the implications of the use of biofuels for global security as well as for economic, environmental, and public health need to be further evaluated".

According to Francisco Blanch, a commodity strategist for Merrill Lynch, crude oil would be trading 15 per cent higher and gasoline would be as much as 25 per cent more expensive, if it were not for biofuels. Gordon Quaiattini, president of the Canadian Renewable Fuels Association, argued that a healthy supply of alternative energy sources will help to combat gasoline price spikes.

"Food vs. Fuel" Debate

Food vs fuel is the debate regarding the risk of diverting farmland or crops for biofuels production in detriment of the food supply on a global scale. Essentially the debate refers to the possibility that by farmers increasing their production of these crops, often through government subsidy incentives, their time and land is shifted away from other types of non-biofuel crops driving up the price of non-biofuel crops due to the decrease in production. Therefore, it is not only that there is an increase in demand for the food staples, like corn and cassava, that sustain the majority of the world's poor but this also has the potential to increase the price of the remaining crops that these individuals would otherwise need to utilize to supplement their diets. A recent study for the International Centre for Trade and Sustainable Development shows that market-driven expansion of ethanol in the US increased maize prices by 21 percent in 2009, in comparison with what prices would have been had ethanol production been frozen at 2004 levels. A November 2011 study states that biofuels, their production, and their subsidies are leading causes of agricultural price shocks. The counter-argument includes considerations of the type of corn that is utilized in biofuels, often field corn not suitable for human consumption; the portion of the corn that is used in ethanol, the starch portion; and the negative effect higher prices for corn and grains have on government welfare for these products. The "food vs. fuel" or "food or fuel" debate is internationally controversial, with disagreement about how significant this is, what is causing it, what the effect is, and what can or should be done about it.

Poverty Reduction Potential

Researchers at the Overseas Development Institute have argued that biofuels could help to reduce poverty in the developing world, through increased employment, wider economic growth multipliers and by stabilising oil prices (many developing countries are net importers of oil). However, this potential is described as 'fragile', and is reduced where feedstock production tends to be large scale, or causes pressure on limited agricultural resources: capital investment, land, water, and the net cost of food for the poor.

With regards to the potential for poverty reduction or exacerbation, biofuels rely on many of the same policy, regulatory or investment shortcomings that impede agriculture as a route to poverty reduction. Since many of these shortcomings require policy improvements at a country level rather than a global one, they argue for a country-by-country analysis of the potential poverty effects of biofuels. This would consider, among other things, land administration systems, market coordination and prioritizing investment in biodiesel, as this 'generates more labour, has lower transportation costs and uses simpler technology'. Also necessary are reductions in the tariffs on biofuel imports regardless of the country of origin, especially due to the increased efficiency of biofuel production in countries such as Brazil.

Sustainable Biofuel Production

Responsible policies and economic instruments would help to ensure that biofuel commercialization, including the development of new cellulosic technologies, is sustainable. Responsible commercialization of biofuels represents an opportunity to enhance sustainable economic prospects in Africa, Latin America and impoverished Asia.

Environmental Effects

Soil Erosion and Deforestation

Large-scale deforestation of mature trees (which help remove CO_2 through photosynthesis — much better than sugar cane or most other biofuel feedstock crops do) contributes to soil erosion, unsustainable global warming atmospheric greenhouse gas levels, loss of habitat, and a reduction of valuable biodiversity (both on land as in oceans). Demand for biofuel has led to clearing land for palm oil plantations. In Indonesia alone, over 9,400,000 acres (38,000 km²) of forest have been converted to plantations since 1996.

A portion of the biomass should be retained onsite to support the soil resource. Normally this will be in the form of raw biomass, but processed biomass is also an option. If the exported biomass is used to produce syngas, the process can be used to co-produce biochar, a low-temperature charcoal used as a soil amendment to increase soil organic matter to a degree not practical with less recalcitrant forms of organic carbon. For co-production of biochar to be widely adopted, the soil amendment and carbon sequestration value of co-produced charcoal must exceed its net value as a source of energy.

Some commentators claim that removal of additional cellulosic biomass for biofuel production will further deplete soils.

Effect on Water Resources

Increased use of biofuels puts increasing pressure on water resources in at least two ways: water use for the irrigation of crops used as feedstocks for biodiesel production; and water use in the production of biofuels in refineries, mostly for boiling and cooling.

In many parts of the world supplemental or full irrigation is needed to grow feedstocks. For example, if in the production of corn (maize) half the water needs of crops are met through irrigation and the other half through rainfall, about 860 liters of water are needed to produce one liter of ethanol. However, in the United States only 5-15% of the water required for corn comes from irrigation while the other 85-95% comes from natural rainfall.

In the United States, the number of ethanol factories has almost tripled from 50 in 2000 to about 140 in 2008. A further 60 or so are under construction, and many more are planned. Projects are being challenged by residents at courts in Missouri (where water is drawn from the Ozark Aquifer), Iowa, Nebraska, Kansas (all of which draw water from the non-renewable Ogallala Aquifer), central Illinois (where water is drawn from the Mahomet Aquifer) and Minnesota.

For example, the four ethanol crops: corn, sugarcane, sweet sorghum and pine yield net energy. However, increasing production in order to meet the U.S. Energy Independence and Security Act mandates for renewable fuels by 2022 would take a heavy toll in the states of Florida and Georgia. The sweet sorghum, which performed the best of the four, would increase the amount of freshwater withdrawals from the two states by almost 25%.

Loss of Biodiversity

Critics argue that expansion of farming for biofuel production causes unacceptable loss of biodiversity for a much less significant decrease in fossil fuel consumption. The loss of biodiversity also makes heavy dependence on biofuels very risky by reducing our ability to deal with blights affecting the few important biofuel crops. Food crops have recovered from blights when the old stock was mixed with blight resistant wild strains, but as biodiversity is lost due to excessive agriculture, the possibilities for recovering from blights are lost.

Pollution

Formaldehyde, acetaldehyde and other aldehydes are produced when alcohols are oxidized. When only a 10% mixture of ethanol is added to gasoline (as is common in American E10 gasohol and elsewhere), aldehyde emissions increase 40%. Some study results are conflicting on this fact however, and lowering the sulfur content of biofuel mixes lowers the acetaldehyde levels. Burning biodiesel also emits aldehydes and other potentially hazardous aromatic compounds which are not regulated in emissions laws.

Many aldehydes are toxic to living cells. Formaldehyde irreversibly cross-links protein amino acids, which produces the hard flesh of embalmed bodies. At high concentrations in an enclosed space, formaldehyde can be a significant respiratory irritant causing nose bleeds, respiratory distress, lung disease, and persistent headaches. Acetaldehyde, which is produced in the body by alcohol drinkers and found in the mouths of smokers and those with poor oral hygiene, is carcinogenic and mutagenic.

The European Union has banned products that contain Formaldehyde, due to its documented carcinogenic characteristics. The U.S. Environmental Protection Agency has labeled Formaldehyde as a probable cause of cancer in humans.

Brazil burns significant amounts of ethanol biofuel. Gas chromatograph studies were performed of ambient air in São Paulo, Brazil, and compared to Osaka, Japan, which does not burn ethanol fuel. Atmospheric Formaldehyde was 160% higher in Brazil, and Acetaldehyde was 260% higher.

Technical Issues

Energy Efficiency and Energy Balance

Production of biofuels from raw materials requires energy (for farming, transport and conversion to final product, and the production / application of fertilizers, pesticides, herbicides, and fungicides), and has environmental consequences.

The energy balance of a biofuel (sometimes called "Net energy gain" and EROEI) is determined by the amount of energy put into the manufacture of fuel compared to the amount of energy released when it is burned in a vehicle. This varies by feedstock and according to the assumptions used. Biodiesel made from sunflowers may produce only 0.46 times the input rate of fuel energy. Biodiesel made from soybeans may produce 3.2 times the input rate of fossil fuels. This compares to 0.805 for gasoline and 0.843 for diesel made from petroleum. Biofuels may require higher energy input per unit of BTU energy content produced than fossil fuels: petroleum can be pumped out of the ground and processed more efficiently than biofuels can be grown and processed. However, this is not necessarily a reason to use oil instead of biofuels, nor does it affect the environmental benefits provided by a given biofuel.

Studies have been done that calculate energy balances for biofuel production. Some of these show large differences depending on the biomass feedstock used and location.

To explain one specific example, a June 17, 2006 editorial in the Wall. St. Journal stated, "The most widely cited research on this subject comes from Cornell's David Pimental and Berkeley's Ted Patzek. They've found that it takes more than a gallon of fossil fuel to make one gallon of ethanol — 29% more. That's because it takes enormous amounts of fossil-fuel energy to grow corn (using fertilizer and irrigation), to transport the crops and then to turn that corn into ethanol."

Life cycle assessments of biofuel production show that under certain circumstances, biofuels produce only limited savings in energy and greenhouse gas emissions. Fertilizer inputs and transportation of biomass across large distances can reduce the greenhouse gas (GHG) savings achieved. The location of biofuel processing plants can be planned to minimize the need for transport, and agricultural regimes can be developed to limit the amount of fertiliser used for biomass production. A European study on the greenhouse gas emissions found that well-to-wheel (WTW) CO_2 emissions of biodiesel from seed crops such as rapeseed could be almost as high as fossil diesel. It showed a similar result for bio-ethanol from starch crops, which could have almost as many WTW CO_2 emissions as fossil petrol. This study showed that second generation biofuels have far lower WTW CO_2 emissions.

Other independent LCA studies show that biofuels save around 50% of the CO_2 emissions of the equivalent fossil fuels. This can be increased to 80-90% GHG emissions savings if second generation processes or reduced fertiliser growing regimes are used. Further GHG savings can be achieved by using by-products to provide heat, such as using bagasse to power ethanol production from sugarcane.

Collocation of synergistic processing plants can enhance efficiency. One example is to use the exhaust heat from an industrial process for ethanol production, which can then recycle cooler processing water, instead of evaporating hot water that warms the atmosphere.

Biomass planting mandated by law (as in European Union) results in large quantities of biomass being transported to EU from Africa, Asia and Americas (Canada, USA, Brazil). For example, in Poland as much as 85% of biomass used is imported from outside of EU, with single electric plant in Łódź importing over 7'000 tons of wood biomass from Republic of Komi (Russia) over distance of 7'000 kilometers on monthly basis.

T. A. Kiefer of the US Air Force Air War College, in a paper entitled "The False Promise of Liquid Biofuels", laid out factors that preclude biofuels from replacing petroleum as a national-scale transportation fuel. Kiefer states "The energy content of the final-product biofuel compared to the energy required to produce it proves to be a very poor investment, especially compared to other alternatives. In many cases, there is net loss of energy." He concludes "...pursuit of biofuels creates irreversible harm to the environment, increases greenhouse gas emissions, undermines food security, and promotes abuse of human rights."

Solar Energy Efficiency

Biofuels from plant materials convert energy that was originally captured from solar energy via photosynthesis. A comparison of conversion efficiency from solar to usable energy (taking into account the whole energy budgets) shows that photovoltaics are 100 times more efficient than corn ethanol and 10 times more efficient than the best biofuel. However, photovoltaics produce electricity rather than storable, portable liquid hydrocarbon fuel, so they are largely irrelevant for powering the large existing fleet of vehicles and equipment having internal combustion engines. Also from the economic point of view, green plants are self-assembling organisms and therefore much cheaper to produce than photovoltaic cells.

Carbon Emissions

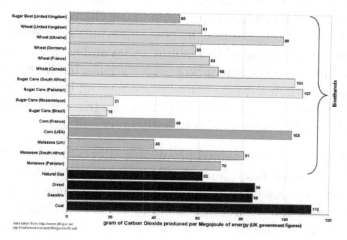

Graph of UK figures for the carbon intensity of bioethanol and fossil fuels. This graph assumes that all bioethanols are burnt in their country of origin and that previously existing cropland is used to grow the feedstock.

Biofuels and other forms of renewable energy aim to be carbon neutral or even carbon negative. Carbon neutral means that the carbon released during the use of the fuel, e.g. through burning to power transport or generate electricity, is reabsorbed and balanced by the carbon absorbed by new plant growth. These plants are then harvested to make the next batch of fuel. Carbon neutral fuels lead to no net increases in human contributions to atmospheric carbon dioxide levels, reducing the human contributions to global warming. A carbon negative aim is achieved when a portion of the biomass is used for carbon sequestration. Calculating exactly how much greenhouse gas (GHG) is produced in burning biofuels is a complex and inexact process, which depends very much on the method by which the fuel is produced and other assumptions made in the calculation.

The carbon emissions (carbon footprint) produced by biofuels are calculated using a technique called Life Cycle Analysis (LCA). This uses a "cradle to grave" or "well to wheels" approach to calculate the total amount of carbon dioxide and other greenhouse gases emitted during biofuel production, from putting seed in the ground to using the fuel in cars and trucks. Many different LCAs have been done for different biofuels, with widely differing results. Several well-to-wheel analysis for biofuels has shown that first generation biofuels can reduce carbon emissions, with savings depending on the feedstock used, and second generation biofuels can produce even higher savings when compared to using fossil fuels. However, those studies did not take into account emissions from nitrogen fixation, or additional carbon emissions due to indirect land use changes. In addition, many LCA studies fail to analyze the effect of substitutes that may come into the market to replace current biomass-based products. In the case of Crude Tall Oil, a raw material used in the production of pine chemicals and now being diverted for use in biofuel, an LCA study found that the global carbon footprint of pine chemicals produced from CTO is 50 percent lower than substitute products used in the same situation offsetting any gains from utilizing a biofuel to replace fossil fuels. Additionally the study showed that fossil fuels are not reduced when CTO is diverted to biofuel use and the substitute products consume disproportionately more energy. This diversion will negatively affect an industry that contributes significantly to the world economy, globally producing more than 3 billion pounds of pine chemicals annually in complex, high technology refineries and providing jobs directly and indirectly for tens of thousands of workers.

A paper published in February 2008 in Sciencexpress by a team led by Searchinger from Princeton University concluded that once considered indirect land use changes effects in the life cycle assessment of biofuels used to substitute gasoline, instead of savings both corn and cellulosic ethanol increased carbon emissions as compared to gasoline by 93 and 50 percent respectively. A second paper published in the same issue of Sciencexpress, by a team led by Fargione from The Nature Conservancy, found that a carbon debt is created when natural lands are cleared and being converted to biofuel production and to crop production when agricultural land is diverted to biofuel production, therefore this carbon debt applies to both direct and indirect land use changes.

The Searchinger and Fargione studies gained prominent attention in both the popular media and in scientific journals. The methodology, however, drew some criticism, with Wang and Haq from Argonne National Laboratory posted a public letter and send their criticism about the Searchinger paper to Letters to Science. Another criticism by Kline and Dale from Oak Ridge National Laboratory was published in Letters to Science. They argued that Searchinger et al. and Fargione et al. "...*do not provide adequate support for their claim that biofuels cause high emissions due to land-use change.* The U.S. biofuel industry also reacted, claiming in a public letter, that the "*Searchinger study is clearly a "worst case scenario" analysis...*" and that this study "*relies on a long series of highly subjective assumptions...*".

Modifications Necessary to Internal Combustion Engines

The modifications necessary to run internal combustion engines on biofuel depend on the type of biofuel used, as well as the type of engine used. For example, gasoline engines can run without any modification at all on biobutanol. Minor modifications are however needed to run on bioethanol or biomethanol. Diesel engines can run on the latter fuels, as well as on vegetable oils (which are cheaper). However, the latter is only possible when the engine has been foreseen with indirect injection. If no indirect injection is present, the engine hence needs to be fitted with this.

Campaigns

A number of environmental NGOs campaign against the production of biofuels as a large-scale alternative to fossil fuels. For example, Friends of the Earth state that "the current rush to develop agrofuels (or biofuels) on a large scale is ill-conceived and will contribute to an already unsustainable trade whilst not solving the problems of climate change or energy security". Some mainstream environmental groups support biofuels as a significant step toward slowing or stopping global climate change. However, supportive environmental groups generally hold the view that biofuel production can threaten the environment if it is not done sustainably. This finding has been backed by reports of the UN, the IPCC, and some other smaller environmental and social groups as the EEB and the Bank Sarasin, which generally remain negative about biofuels.

As a result, governmental and environmental organizations are turning against biofuels made in a non-sustainable way (hereby preferring certain oil sources as jatropha and lignocellulose over palm oil) and are asking for global support for this. Also, besides supporting these more sustainable biofuels, environmental organizations are redirecting to new technologies that do not use internal combustion engines such as hydrogen and compressed air.

Several standard-setting and certification initiatives have been set up on the topic of biofuels. The "Roundtable on Sustainable Biofuels" is an international initiative which brings together farmers, companies, governments, non-governmental organizations, and scientists who are interested in the sustainability of biofuels production and distribution. During 2008, the Roundtable is developing a series of principles and criteria for sustainable biofuels production through meetings, teleconferences, and online discussions. In a similar vein, the Bonsucro standard has been developed as a metric-based certificate for products and supply chains, as a result of an ongoing multi-stakeholder initiative focussing on the products of sugar cane, including ethanol fuel.

The increased manufacture of biofuels will require increasing land areas to be used for agriculture. Second and third generation biofuel processes can ease the pressure on land, because they can use waste biomass, and existing (untapped) sources of biomass such as crop residues and potentially even marine algae.

In some regions of the world, a combination of increasing demand for food, and increasing demand for biofuel, is causing deforestation and threats to biodiversity. The best reported example of this is the expansion of oil palm plantations in Malaysia and Indonesia, where rainforest is being destroyed to establish new oil palm plantations. It is an important fact that 90% of the palm oil produced in Malaysia is used by the food industry; therefore biofuels cannot be held solely responsible for this deforestation. There is a pressing need for sustainable palm oil production for the food and fuel industries; palm oil is used in a wide variety of food products. The *Roundtable on Sustainable Biofuels* is working to define criteria, standards and processes to promote sustainably produced biofuels. Palm oil is also used in the manufacture of detergents, and in electricity and heat generation both in Asia and around the world (the UK burns palm oil in coal-fired power stations to generate electricity).

Significant area is likely to be dedicated to sugar cane in future years as demand for ethanol increases worldwide. The expansion of sugar cane plantations will place pressure on environmentally sensitive native ecosystems including rainforest in South America. In forest ecosystems, these effects themselves will undermine the climate benefits of alternative fuels, in addition to representing a major threat to global biodiversity.

Although biofuels are generally considered to improve net carbon output, biodiesel and other fuels do produce local air pollution, including nitrogen oxides, the principal cause of smog.

Politics

Steven Rattner, former "auto czar" for U.S. President Barack Obama, wrote an Op-ed for *The New York Times* in June, 2011, entitled "The Great Corn Con," characterizing ethanol as "an example of government policy run amok." Along with the economic and environmental effects of the U.S. policy, he noted the effect of the issue on presidential politics:

Those [presidential] hopefuls have seen no need for a foolish consistency. John McCain and John Kerry were against ethanol subsidies, then as candidates were for them. Having lost the presidency, Mr. McCain is now against them again. Al Gore was for ethanol before he was against it. This time, one hopeful is experimenting with counter-programming: as governor of corn-producing Minnesota, Tim Pawlenty pushed for subsidies before he embraced a "straight talk" strategy.

Rattner did not address President Obama's long-time alignment with Illinois and U.S. corn producers on the issue.

In April, 2014 an article in the Financial Times described how biofuel manufacturers in Europe felt threatened by changing European Union climate policies. After mapping rigorous regulations for cleaner fuels in 2009, the European Commission decided not to set specific fuel targets from 2020 to 2030. Instead the Commission recommended a "more holistic and integrated approach" to creating an efficient biofuels policy. The EC also called for "an improved biomass policy" to "maximize the resource efficient use of biomass in order to deliver robust and verifiable greenhouse gas savings and to allow for fair competition between the various uses of biomass resources in the construction sector, paper and pulp industries and biochemical and energy production." Fair competition in the acquisition of biomass feedstocks is what established bio-industries are asking for. Incentives provided to biofuel producers create an uneven playing field. "The pine chemical industry, which uses a co product called crude tall oil from the paper-pulping process, says it's concerned that incentives will divert its major raw material into biofuel production," stated a Greenwire article.

Food Vs. Fuel

An ethanol fuel plant under construction, Butler County, Iowa

Food versus fuel is the dilemma regarding the risk of diverting farmland or crops for biofuels production to the detriment of the food supply. The biofuel and food price debate involves wide-ranging views, and is a long-standing, controversial one in the literature. There is disagreement about the significance of the issue, what is causing it, and what can or should be done to remedy the situation. This complexity and uncertainty is due to the large number of impacts and feedback loops that can positively or negatively affect the price system. Moreover, the relative strengths of these positive and negative impacts vary in the short and long terms, and involve delayed effects. The academic side of the debate is also blurred by the use of different economic models and competing forms of statistical analysis.

Biofuel production has increased in recent years. Some commodities like maize (corn), sugar cane or vegetable oil can be used either as food, feed, or to make biofuels. For example, since 2006, a portion of land that was also formerly used to grow other crops in the United States is now used to grow corn for biofuels, and a larger share of corn is destined to ethanol production, reaching 25% in 2007. Second generation biofuels could potentially combine farming for food and fuel and moreover, electricity could be generated simultaneously, which could be beneficial for developing countries and rural areas in developed countries. With global demand for biofuels on the increase due to the oil price increases taking place since 2003 and the desire to reduce oil dependency as well as reduce GHG emissions from transportation, there is also fear of the potential destruction of natural habitats by being converted into farmland. Environmental groups have raised concerns about this trade-off for several years, but now the debate reached a global scale due to the 2007–2008 world food price crisis. On the other hand, several studies do show that biofuel production can be significantly increased without increased acreage. Therefore, stating that the crisis in hand relies on the food scarcity.

Biofuels are not a new phenomenon. Before the industrialisation, horses were the primary (and humans probably the secondary) source of power for transportation and physical work, requiring food. The growing of crops for horses (typically oat) for carrying out physical work is of course comparable to the growing of crops for biofuels for engines, albeit on a smaller scale, because production since then has increased.

Brazil has been considered to have the world's first sustainable biofuels economy and its government claims Brazil's sugar cane based ethanol industry has not contributed to the 2008 food crisis. A World Bank policy research working paper released in July 2008 concluded that "...large increases in biofuels production in the United States and Europe are the main reason behind the steep rise in global food prices", and also stated that "Brazil's sugar-based ethanol did not push food prices appreciably higher". However, a 2010 study also by the World Bank concluded that their previous study may have overestimated the contribution of biofuel production, as "the effect of biofuels on food prices has not been as large as originally thought, but that the use of commodities by financial investors (the so-called "financialisation of commodities") may have been partly responsible for the 2007/08 spike." A 2008 independent study by OECD also found that the impact of biofuels on food prices are much smaller.

Food Price Inflation

From 1974 to 2005 real food prices (adjusted for inflation) dropped by 75%. Food commodity prices were relatively stable after reaching lows in 2000 and 2001. Therefore, recent rapid food price increases are considered extraordinary. A World Bank policy research working paper published on July 2008 found that the increase in food commodities prices was led by grains, with sharp price increases in 2005 despite record crops worldwide. From January 2005 until June 2008, maize prices almost tripled, wheat increased 127 percent, and rice rose 170 percent. The increase in grain prices was followed by increases in fats and oil prices in mid-2006. On the other hand, the study found that sugar cane production has increased rapidly, and it was large enough to keep sugar price increases small except for 2005 and early 2006. The paper concluded that biofuels produced from grains have raised food prices in combination with other related factors between 70 to 75 percent, but ethanol produced from sugar cane has not contributed significantly to the recent increase in food commodities prices.

An economic assessment report published by the OECD in July 2008 found that "...the impact of current biofuel policies on world crop prices, largely through increased demand for cereals and vegetable oils, is significant but should not be overestimated. Current biofuel support measures alone are estimated to increase average wheat prices by about 5 percent, maize by around 7 percent and vegetable oil by about 19 percent over the next 10 years."

Corn is used to make ethanol and prices went up by a factor of three in less than 3 years (measured in US dollars). Reports in 2007 linked stories as diverse as food riots in Mexico due to rising prices of corn for tortillas, and reduced profits at Heineken the large international brewer, to the increasing use of corn (maize) grown in the US Midwest for ethanol production. (In the case of beer, the barley area was cut in order to increase corn production. Barley is not currently used to produce ethanol.) Wheat is up by almost a factor of 3 in 3 years, while soybeans are up by a factor of 2 in 2 years (both measured in US dollars).

As corn is commonly used as feed for livestock, higher corn prices lead to higher prices in animal source foods. Vegetable oil is used to make biodiesel and has about doubled in price in the last couple years. The price is roughly tracking crude oil prices. The 2007–2008 world food price crisis is blamed partly on the increased demand for biofuels. During the same period rice prices went up by a factor of 3 even though rice is not directly used in biofuels.

The USDA expects the 2008/2009 wheat season to be a record crop and 8% higher than the previous year. They also expect rice to have a record crop. Wheat prices have dropped from a high over $12/bushel in May 2008 to under $8/bushel in May. Rice has also dropped from its highs.

According to a 2008 report from the World Bank the production of biofuel pushed food prices up. These conclusions were supported by the Union of Concerned Scientists in their September 2008 newsletter in which they remarked that the World Bank analysis "contradicts U.S. Secretary of Agriculture Ed Schaffer's assertion that biofuels account for only a small percentage of rising food prices."

According to the October Consumer Price Index released Nov. 19, 2008, food prices continued to rise in October 2008 and were 6.3 percent higher than October 2007. Since July 2008 fuel costs dropped by nearly 60 percent.

Proposed Causes

Ethanol Fuel as an Oxygenate Additive

The demand for ethanol fuel produced from field corn was spurred in the U.S. by the discovery that methyl tertiary butyl ether (MTBE) was contaminating groundwater. MTBE use as an oxygenate additive was widespread due to mandates of the Clean Air Act amendments of 1992 to reduce carbon monoxide emissions. As a result, by 2006 MTBE use in gasoline was banned in almost 20 states. There was also concern that widespread and costly litigation might be taken against the U.S. gasoline suppliers, and a 2005 decision refusing legal protection for MTBE, opened a new market for ethanol fuel, the primary substitute for MTBE. At a time when corn prices were around US$2 a bushel, corn growers recognized the potential of this new market and delivered accordingly. This demand shift took place at a time when oil prices were already significantly rising.

Other Factors

That food prices went up at the same time fuel prices went up is not surprising and should not be entirely blamed on biofuels. Energy costs are a significant cost for fertilizer, farming, and food distribution. Also, China and other countries have had significant increases in their imports as their economies have grown. Sugar is one of the main feedstocks for ethanol and prices are down from 2 years ago. Part of the food price increase for international food commodities measured in US dollars is due to the dollar being devalued. Protectionism is also an important contributor to price increases. 36% of world grain goes as fodder to feed animals, rather than people.

Over long time periods population growth and climate change could cause food prices to go up. However, these factors have been around for many years and food prices have jumped up in the last 3 years, so their contribution to the current problem is minimal.

Government Regulations of Food and Fuel Markets

France, Germany, the United Kingdom and the United States governments have supported biofuels with tax breaks, mandated use, and subsidies. These policies have the unintended consequence of diverting resources from food production and leading to surging food prices and the potential destruction of natural habitats.

Fuel for agricultural use often does not have fuel taxes (farmers get duty-free petrol or diesel fuel). Biofuels may have subsidies and low/no retail fuel taxes. Biofuels compete with retail gasoline and diesel prices which have substantial taxes included. The net result is that it is possible for a farmer to use more than a gallon of fuel to make a gallon of biofuel and still make a profit. Some argue that this is a bad distortion of the market. There have been thousands of scholarly papers analyzing how much energy goes into making ethanol from corn and how that compares to the energy in the ethanol.

A World Bank policy research working paper concluded that food prices have risen by 35 to 40 percent between 2002 and 2008, of which 70 to 75 percent is attributable to biofuels. The "month-by-month" five-year analysis disputes that increases in global grain consumption and droughts were responsible for significant price increases, reporting that this had only a marginal impact. Instead the report argues that the EU and US drive for biofuels has had by far the biggest impact on food supply and prices, as increased production of biofuels in the US and EU were supported by subsidies and tariffs on imports, and considers that without these policies, price increases would have been smaller. This research also concluded that Brazil's sugar cane based ethanol has not raised sugar prices significantly, and recommends removing tariffs on ethanol imports by both the US and EU, to allow more efficient producers such as Brazil and other developing countries, including many African countries, to produce ethanol profitably for export to meet the mandates in the EU and the US.

An economic assessment published by the OECD in July 2008 agrees with the World Bank report recommendations regarding the negative effects of subsidies and import tariffs, but found that the estimated impact of biofuels on food prices are much smaller. The OECD study found that trade restrictions, mainly through import tariffs, protect the domestic industry from foreign competitors but impose a cost burden on domestic biofuel users and limits alternative suppliers. The report is also critical of limited reduction of GHG emissions achieved from biofuels based on feedstocks used in Europe and North America, finding that the current biofuel support policies would reduce greenhouse gas emissions from transport fuel by no more than 0.8% by 2015, while Brazilian ethanol from sugar cane reduces greenhouse gas emissions by at least 80% compared to fossil fuels. The assessment calls for the need for more open markets in biofuels and feedstocks in order to improve efficiency and lower costs.

Oil Price Increases

Oil price increases since 2003 resulted in increased demand for biofuels. Transforming vegetable oil into biodiesel is not very hard or costly so there is a profitable arbitrage situation if vegetable oil is much cheaper than diesel. Diesel is also made from crude oil, so vegetable oil prices are partially linked to crude oil prices. Farmers can switch to growing vegetable oil crops if those are more profitable than food crops. So all food prices are linked to vegetable oil prices, and in turn to crude oil prices. A World Bank study concluded that oil prices and a weak dollar explain 25-30% of total price rise between January 2002 until June 2008.

Demand for oil is outstripping the supply of oil and oil depletion is expected to cause crude oil prices to go up over the next 50 years. Record oil prices are inflating food prices worldwide, including those crops that have no relation to biofuels, such as rice and fish.

In Germany and Canada it is now much cheaper to heat a house by burning grain than by using fuel derived from crude oil. With oil at $120/barrel a savings of a factor of 3 on heating costs is possible. When crude oil was at $25/barrel there was no economic incentive to switch to a grain fed heater.

From 1971 to 1973, around the time of the 1973 oil crisis, corn and wheat prices went up by a factor of 3. There was no significant biofuel usage at that time.

US Government Policy

Some argue that the US government policy of encouraging ethanol from corn is the main cause for food price increases. US Federal government ethanol subsidies total $7 billion per year, or $1.90 per gallon. Ethanol provides only 55% as much energy as gasoline per gallon, realizing about a $3.45 per gallon gasoline trade off. Corn is used to feed chickens, cows, and pigs, so higher corn prices lead to higher prices for chicken, beef, pork, milk, cheese, etc.

U.S. Senators introduced the *BioFuels Security Act* in 2006. "It's time for Congress to realize what farmers in America's heartland have known all along - that we have the capacity and ingenuity to decrease our dependence on foreign oil by growing our own fuel," said U.S. Senator for Illinois Barack Obama.

Two-thirds of U.S. oil consumption is due to the transportation sector. The Energy Independence and Security Act of 2007 has a significant impact on U.S. Energy Policy. With the high profitability of growing corn, more and more farmers switch to growing corn until the profitability of other crops goes up to match that of corn. So the ethanol/corn subsidies drive up the prices of other farm crops.

The US - an important export country for food stocks - will convert 18% of its grain output to ethanol in 2008. Across the US, 25% of the whole corn crop went to ethanol in 2007. The percentage of corn going to biofuel is expected to go up.

Since 2004 a US subsidy has been paid to companies that blend biofuel and regular fuel. The European biofuel subsidy is paid at the point of sale. Companies import biofuel to the US, blend 1% or even 0.1% regular fuel, and then ship the blended fuel to Europe, where it can get a second subsidy. These blends are called B99 or B99.9 fuel. The practice is called "splash and dash". The imported fuel may even come from Europe to the US, get 0.1% regular fuel, and then go back to Europe. For B99.9 fuel the US blender gets a subsidy of $0.999 per gallon. The European biodiesel producers have urged the EU to impose punitive duties on these subsidized imports. In 2007, US lawmakers were also looking at closing this loophole.

Freeze on First Generation Biofuel Production

The prospects for the use of biofuels could change in a relatively dramatic way in 2014. Petroleum trade groups petitioned the EPA in August 2013 to take into consideration a reduction of renewable biofuel content in transportation fuels. On November 15, 2013 the United States EPA announced a review of the proportion of ethanol that should be required by regulation. The standards established by the Energy Independence and Security Act of 2007 could be modified significantly. The announcement allows sixty days for the submission of commentary about the proposal.

Journalist George Monbiot has argued for a 5-year freeze on biofuels while their impact on poor communities and the environment is assessed. It has been suggested that a problem with Monbiot's approach is that economic drivers may be required in order to push through the development of more sustainable second-generation biofuel processes: it is possible that these could be stalled if biofuel production decreases. Some environmentalists are suspicious that second-generation biofuels may not solve the problem of a potential clash with food as they also use significant agricultural resources such as water.

A recent UN report on biofuel also raises issues regarding food security and biofuel production. Jean Ziegler, then UN Special Rapporteur on food, concluded that while the argument for biofuels in terms of energy efficiency and climate change are legitimate, the effects for the world's hungry of transforming wheat and maize crops into biofuel are "absolutely catastrophic," and terms such use of arable land a "crime against humanity." Ziegler also calls for a 5-year moratorium on biofuel production. Ziegler's proposal for a five-year ban was rejected by the U.N. Secretary Ban Ki-moon, who called for a comprehensive review of the policies on biofuels, and said that "just criticising biofuel may not be a good solution".

Food surpluses exist in many developed countries. For example, the UK wheat surplus was around 2 million tonnes in 2005. This surplus alone could produce sufficient bioethanol to replace around 2.5% of the UK's petroleum consumption, without requiring any increase in wheat cultivation or reduction in food supply or exports. However, above a few percent, there would be direct competition between first generation biofuel production and food production. This is one reason why many view second generation biofuels as increasingly important.

Non-food Crops for Biofuel

There are different types of biofuels and different feedstocks for them, and it has been proposed that only non-food crops be used for biofuel. This avoids direct competition for commodities like corn and edible vegetable oil. However, as long as farmers are able to derive a greater profit by switching to biofuels, they will. The law of supply and demand predicts that if fewer farmers are producing food the price of food will rise.

Second generation biofuels use lignocellulosic raw material such as forest residues (sometimes referred to as brown waste and black liquor from Kraft process or sulfite process pulp mills). Third generation biofuels (biofuel from algae) use non-edible raw materials sources that can be used for biodiesel and bioethanol.

Biodiesel

Soybean oil, which only represents half of the domestic raw materials available for biodiesel production in the United States, is one of many raw materials that can be used to produce biodiesel.

Non-food crops like Camelina, Jatropha, seashore mallow and mustard, used for biodiesel, can thrive on marginal agricultural land where many trees and crops won't grow, or would produce only slow growth yields. Camelina is virtually 100 percent efficient. It can be harvested and crushed for oil and the remaining parts can be used to produce high quality omega-3 rich animal feed, fiberboard, and glycerin. Camelina does not take away from land currently being utilized for food production. Most camelina acres are grown in areas that were previously not utilized for farming. For example, areas that receive limited rainfall that can not sustain corn or soybeans without the addition of irrigation can grow camelina and add to their profitability.

Jatropha cultivation provides benefits for local communities:

Cultivation and fruit picking by hand is labour-intensive and needs around one person per hectare. In parts of rural India and Africa this provides much-needed jobs - about 200,000 people worldwide now find employment through jatropha. Moreover, villagers often find that they can grow other crops in the shade of the trees. Their communities will avoid importing expensive diesel and there will be some for export too.

NBB's Feedstock Development program is addressing production of arid variety crops, algae, waste greases, and other feedstocks on the horizon to expand available material for biodiesel in a sustainable manner.

Bioalcohols

Cellulosic ethanol is a type of biofuel produced from lignocellulose, a material that comprises much of the mass of plants. Corn stover, switchgrass, miscanthus and woodchip are some of the more popular non-edible cellulosic materials for ethanol production. Commercial investment in such second-generation biofuels began in 2006/2007, and much of this investment went beyond pilot-scale plants. Cellulosic ethanol commercialization is moving forward rapidly. The world's first commercial wood-to-ethanol plant began operation in Japan in 2007, with a capacity of 1.4 million liters/year. The first wood-to-ethanol plant in the United States is planned for 2008 with an initial output of 75 million liters/year.

Other second generation biofuels may be commercialized in the future and compete less with food. Synthetic fuel can be made from coal or biomass and may be commercialized soon.

Biofuel from Food Byproducts and Coproducts

Biofuels can also be produced from the waste byproducts of food-based agriculture (such as citrus peels or used vegetable oil) to manufacture an environmentally sustainable fuel supply, and reduce waste disposal cost.

A growing percentage of U.S. biodiesel production is made from waste vegetable oil (recycled restaurant oils) and greases.

Collocation of a waste generator with a waste-to-ethanol plant can reduce the waste producer's operating cost, while creating a more-profitable ethanol production business. This innovative collocation concept is sometimes called holistic systems engineering. Collocation disposal elimination may be one of the few cost-effective, environmentally sound, biofuel strategies, but its scalability is limited by availability of appropriate waste generation sources. For example, millions of tons of wet Florida-and-California citrus peels cannot supply billions of gallons of biofuels. Due to the higher cost of transporting ethanol, it is a local partial solution, at best.

More firms are investigating the potential of fractionating technology to remove corn germ (i.e. the portion of the corn kernel that contains oil) prior to the ethanol process. Furthermore, some ethanol plants have already announced their intention to employ technology to remove the remaining vegetable oil from dried distillers grains, a coproduct of the ethanol process. Both of these technologies would add to the biodiesel raw material supply.

Biofuel Subsidies and Tariffs

Some people have claimed that ending subsidies and tariffs would enable sustainable development of a global biofuels market. Taxing biofuel imports while letting petroleum in duty-free does not fit with the goal of encouraging biofuels. Ending mandates, subsidies, and tariffs would end the distortions that current policy is causing. Some US senators advocate reducing subsidies for corn based ethanol. The US ethanol tariff and some US ethanol subsidies are currently set to expire over the next couple years. The EU is rethinking their biofuels directive due to environmental and social concerns. On January 18, 2008 the UK House of Commons Environmental Audit Committee raised similar concerns, and called for a moratorium on biofuel targets. Germany ended their subsidy of biodiesel on Jan 1 2008 and started taxing it.

Reduce Farmland Reserves and Set Asides

To avoid overproduction and to prop up farmgate prices for agricultural commodities, the EU has for a long time have had farm subsidy programs to encourage farmers not to produce and leave productive acres fallow. The 2008 crisis prompted proposals to bring some of the reserve farmland back into use, and the used area increased actually with 0.5% but today these areas are once again out of use. According to Eurostat, 18 million hectares has been abandoned since 1990, 7,4 millions hectares are currently set aside Landuse in EU and the EU has recently decided to set aside another 5-7 %, in so called Ecological Focus Areas, corresponding to 10-12 million hectares. In spite of this reduction of used land, the EU is a net exporter of e.g. wheat.

The American Bakers Association has proposed reducing the amount of farmland held in the US Conservation Reserve Program. Currently the US has 34,500,000 acres (140,000 km²) in the program.

In Europe about 8% of the farmland is in set aside programs. Farmers have proposed freeing up all of this for farming. Two-thirds of the farmers who were on these programs in the UK are not renewing when their term expires.

Sustainable Production of Biofuels

Second generation biofuels are now being produced from the cellulose in dedicated energy crops (such as perennial grasses), forestry materials, the co-products from food production, and domestic vegetable waste. Advances in the conversion processes will almost certainly improve the sustainability of biofuels, through better efficiencies and reduced environmental impact of producing biofuels, from both existing food crops and from cellulosic sources.

Lord Ron Oxburgh suggests that responsible production of biofuels has several advantages:

Produced responsibly they are a sustainable energy source that need not divert any land from growing food nor damage the environment; they can also help solve the problems of the waste generated by Western society; and they can create jobs for the poor where previously were none. Produced irresponsibly, they at best offer no climate benefit and, at worst, have detrimental social and environmental consequences. In other words, biofuels are pretty much like any other product.

Far from creating food shortages, responsible production and distribution of biofuels represents the best opportunity for sustainable economic prospects in Africa, Latin America and impoverished Asia. Biofuels offer the prospect of real market competition and oil price moderation. Crude oil would be trading 15 per cent higher and gasoline would be as much as 25 per cent more expensive, if it were not for biofuels. A healthy supply of alternative energy sources will help to combat gasoline price spikes.

Continuation of the Status Quo

An additional policy option is to continue the current trends of government incentive for these types of crops to further evaluate the effects on food prices over a longer period of time due to the relatively recent onset of the biofuel production industry. Additionally, by virtue of the newness of the industry we can assume that like other startup industries techniques and alternatives will be cultivated quickly if there is sufficient demand for the alternative fuels and biofuels. What could result from the shock to food prices is a very quick move toward some of the non-food biofuels as are listed above amongst the other policy alternatives.

Impact on Developing Countries

Demand for fuel in rich countries is now competing against demand for food in poor countries. The increase in world grain consumption in 2006 happened due to the increase in consumption for fuel, not human consumption. The grain required to fill a 25 US gallons (95 L) fuel tank with ethanol will feed one person for a year.

Several factors combine to make recent grain and oilseed price increases impact poor countries more:

- Poor people buy more grains (e.g. wheat), and are more exposed to grain price changes.

- Poor people spend a higher portion of their income on food, so increasing food prices influence them more.

- Aid organizations which buy food and send it to poor countries see more need when prices go up but are able to buy less food on the same budget.

The impact is not all negative. The Food and Agriculture Organization (FAO) recognizes the potential opportunities that the growing biofuel market offers to small farmers and aquaculturers around the world and has recommended small-scale financing to help farmers in poor countries produce local biofuel.

On the other hand, poor countries that do substantial farming have increased profits due to biofuels. If vegetable oil prices double, the profit margin could more than double. In the past rich countries have been dumping subsidized grains at below cost prices into poor countries and hurting the local farming industries. With biofuels using grains the rich countries no longer have grain surpluses to get rid of. Farming in poor countries is seeing healthier profit margins and expanding.

Interviews with local peasants in southern Ecuador provide strong anecdotal evidence that the high price of corn is encouraging the burning of tropical forests. The destruction of tropical forests now account for 20% of all greenhouse gas emmisons.

National Corn Growers Association

US government subsidies for making ethanol from corn have been attacked as the main cause of the food vs fuel problem. To defend themselves, the National Corn Growers Association has published their views on this issue. They consider the "food vs fuel" argument to be a fallacy that is "fraught with misguided logic, hyperbole and scare tactics."

Claims made by the NCGA include:

- Corn growers have been and will continue to produce enough corn so that supply and demand meet and there is no shortage. Farmers make their planting decisions based on signals from the marketplace. If demand for corn is high and projected revenue-per-acre is strong relative to other crops, farmers will plant more corn. In 2007 US farmers planted 92,900,000 acres (376,000 km²) with corn, 19% more acres than they did in 2006.

- The U.S. has doubled corn yields over the last 40 years and expects to double them again in the next 20 years. With twice as much corn from each acre, corn can be put to new uses without taking food from the hungry or causing deforestation.

- US consumers buy things like corn flakes where the cost of the corn per box is around 5 cents. Most of the cost is packaging, advertising, shipping, etc. Only about 19% of the US retail food prices can be attributed to the actual cost of food inputs like grains and oilseeds. So if the price of a bushel of corn goes up, there may be no noticeable impact on US retail food prices. The US retail food price index has gone up only a few percent per year and is expected to continue to have very small increases.

- Most of the corn produced in the US is field corn, not sweet corn, and not digestible by humans in its raw form. Most corn is used for livestock feed and not human food, even the portion that is exported.

- Only the starch portion of corn kernels is converted to ethanol. The rest (protein, fat, vitamins and minerals) is passed through to the feed coproducts or human food ingredients.

- One of the most significant and immediate benefits of higher grain prices is a dramatic reduction in federal farm support payments. According to the U.S. Department of Agriculture, corn farmers received $8.8 billion in government support in 2006. Because of higher corn prices, payments are expected to drop to $2.1 billion in 2007, a 76 percent reduction.

- While the EROEI and economics of corn based ethanol are a bit weak, it paves the way for cellulosic ethanol which should have much better EROEI and economics.

- While basic nourishment is clearly important, fundamental societal needs of energy, mobility, and energy security are too. If farmers crops can help their country in these areas also, it seems right to do so.

Since reaching record high prices in June 2008, corn prices fell 50% by October 2008, declining sharply together with other commodities, including oil. As ethanol production from corn has continue at the same levels, some have argued that this trend shows the belief that the increased demand for corn to produce ethanol was mistaken. "Analysts, including some in the ethanol sector, say ethanol demand adds about 75 cents to $1.00 per bushel to the price of corn, as a rule of thumb. Other analysts say it adds around 20 percent, or just under 80 cents per bushel at current prices. Those estimates hint that $4 per bushel corn might be priced at only $3 without demand for ethanol fuel.". These industry sources consider that a speculative bubble in the commodity markets holding positions in corn futures was the main driver behind the observed hike in corn prices affecting food supply.

Controversy within the International System

The United States and Brazil lead the industrial world in global ethanol production, with Brazil as the world's largest exporter and biofuel industry leader. In 2006 the U.S. produced 18.4 billion liters (4.86 billion gallons), closely followed by Brazil with 16.3 billion liters (4.3 billion gallons), producing together 70% of the world's ethanol market and nearly 90% of ethanol used as fuel. These countries are followed by China with 7.5%, and India with 3.7% of the global market share.

Since 2007, the concerns, criticisms and controversy surrounding the food vs biofuels issue has reached the international system, mainly heads of states, and inter-governmental organizations (IGOs), such as the United Nations and several of its agencies, particularly the Food and Agriculture

Organization (FAO) and the World Food Programme (WFP); the International Monetary Fund; the World Bank; and agencies within the European Union.

The 2007 Controversy: Ethanol Diplomacy in the Americas

Presidents Luiz Inácio Lula da Silva and George W. Bush during Bush's visit to Brazil, March 2007

In March 2007, "ethanol diplomacy" was the focus of President George W. Bush's Latin American tour, in which he and Brazil's president, Luiz Inácio Lula da Silva, were seeking to promote the production and use of sugar cane based ethanol throughout Latin America and the Caribbean. The two countries also agreed to share technology and set international standards for biofuels. The Brazilian sugar cane technology transfer will permit various Central American countries, such as Honduras, Nicaragua, Costa Rica and Panama, several Caribbean countries, and various Andean Countries tariff-free trade with the U.S. thanks to existing concessionary trade agreements. Even though the U.S. imposes a USD 0.54 tariff on every gallon of imported ethanol, the Caribbean nations and countries in the Central American Free Trade Agreement are exempt from such duties if they produce ethanol from crops grown in their own countries. The expectation is that using Brazilian technology for refining sugar cane based ethanol, such countries could become exporters to the United States in the short-term. In August 2007, Brazil's President toured Mexico and several countries in Central America and the Caribbean to promote Brazilian ethanol technology.

This alliance between the U.S. and Brazil generated some negative reactions. While Bush was in São Paulo as part of the 2007 Latin American tour, Venezuela's President Hugo Chavez, from Buenos Aires, dismissed the ethanol plan as "a crazy thing" and accused the U.S. of trying "to substitute the production of foodstuffs for animals and human beings with the production of foodstuffs for vehicles, to sustain the American way of life." Chavez' complaints were quickly followed by then Cuban President Fidel Castro, who wrote that "you will see how many people among the hungry masses of our planet will no longer consume corn." "Or even worse," he continued, "by offering financing to poor countries to produce ethanol from corn or any other kind of food, no tree will be left to defend humanity from climate change." Daniel Ortega, Nicaragua's President, and one of the preferential recipients of Brazil technical aid, said that "we reject the gibberish of those who applaud Bush's totally absurd proposal, which attacks the food security rights of Latin Americans and Africans, who are major corn consumers", however, he voiced support for sugar cane based ethanol during Lula's visit to Nicaragua.

The 2008 Controversy: Global Food Prices

As a result of the international community's concerns regarding the steep increase in food prices, on April 14, 2008, Jean Ziegler, the United Nations Special Rapporteur on the Right to Food, at the Thirtieth Regional Conference of the Food and Agriculture Organization (FAO) in Brasília, called biofuels a "crime against humanity", a claim he had previously made in October 2007, when he called for a 5-year ban for the conversion of land for the production of biofuels. The previous day, at their Annual International Monetary Fund and World Bank Group meeting at Washington, D.C., the World Bank's President, Robert Zoellick, stated that "While many worry about filling their gas tanks, many others around the world are struggling to fill their stomachs. And it's getting more and more difficult every day."

Luiz Inácio Lula da Silva gave a strong rebuttal, calling both claims "fallacies resulting from commercial interests", and putting the blame instead on U.S. and European agricultural subsidies, and a problem restricted to U.S. ethanol produced from maize. He also said that "biofuels aren't the villain that threatens food security." In the middle of this new wave of criticism, Hugo Chavez reaffirmed his opposition and said that he is concerned that "so much U.S.-produced corn could be used to make biofuel, instead of feeding the world's poor", calling the U.S. initiative to boost ethanol production during a world food crisis a "crime."

German Chancellor Angela Merkel said the rise in food prices is due to poor agricultural policies and changing eating habits in developing nations, not biofuels as some critics claim. On the other hand, British Prime Minister Gordon Brown called for international action and said Britain had to be "selective" in supporting biofuels, and depending on the UK's assessment of biofuels' impact on world food prices, "we will also push for change in EU biofuels targets". Stavros Dimas, European Commissioner for the Environment said through a spokeswoman that "there is no question for now of suspending the target fixed for biofuels", though he acknowledged that the EU had underestimated problems caused by biofuels.

On April 29, 2008, U.S. President George W. Bush declared during a press conference that "85 percent of the world's food prices are caused by weather, increased demand and energy prices", and recognized that "15 percent has been caused by ethanol". He added that "the high price of gasoline is going to spur more investment in ethanol as an alternative to gasoline. And the truth of the matter is it's in our national interests that our farmers grow energy, as opposed to us purchasing energy from parts of the world that are unstable or may not like us." Regarding the effect of agricultural subsidies on rising food prices, Bush said that "Congress is considering a massive, bloated farm bill that would do little to solve the problem. The bill Congress is now considering would fail to eliminate subsidy payments to multi-millionaire farmers", he continued, "this is the right time to reform our nation's farm policies by reducing unnecessary subsidies".

Just a week before this new wave of international controversy began, U.N. Secretary General Ban Ki-moon had commented that several U.N. agencies were conducting a comprehensive review of the policy on biofuels, as the world food price crisis might trigger global instability. He said "We need to be concerned about the possibility of taking land or replacing arable land because of these biofuels", then he added "While I am very much conscious and aware of these problems, at the same time you need to constantly look at having creative sources of energy, including biofuels. Therefore, at this time, just criticising biofuel may not be a good solution. I would urge we need to address these issues in a comprehensive manner." Regarding Jean Ziegler's proposal for a five-year ban, the U.N. Secretary rejected that proposal.

A report released by Oxfam in June 2008 criticized biofuel policies of high-income countries as neither a solution to the climate crisis nor the oil crisis, while contributing to the food price crisis. The report concluded that from all biofuels available in the market, Brazilian sugarcane ethanol is not very effective, but it is the most favorable biofuel in the world in term of cost and greenhouse gas balance. The report discusses some existing problems and potential risks, and asks the Brazilian government for caution to avoid jeopardizing its environmental and social sustainability. The report also says that: "Rich countries spent up to $15 billion last year supporting biofuels while blocking cheaper Brazilian ethanol, which is far less damaging for global food security."

A World Bank research report published on July 2008 found that from June 2002 to June 2008 "biofuels and the related consequences of low grain stocks, large land use shifts, speculative activity and export bans" pushed prices up by 70 percent to 75 percent. The study found that higher oil prices and a weak dollar explain 25-30% of total price rise. The study said that "...large increases in biofuels production in the United States and Europe are the main reason behind the steep rise in global food prices" and also stated that "Brazil's sugar-based ethanol did not push food prices appreciably higher". The Renewable Fuels Association (RFA) published a rebuttal based on the version leaked before its formal release. The RFA critique considers that the analysis is highly subjective and that the author "estimates the impact of global food prices from the weak dollar and the direct and indirect effect of high petroleum prices and attributes everything else to biofuels."

An economic assessment by the OECD also published on July 2008 agrees with the World Bank report regarding the negative effects of subsidies and trade restrictions, but found that the impact of biofuels on food prices are much smaller. The OECD study is also critical of the limited reduction of GHG emissions achieved from biofuels produced in Europe and North America, concluding that the current biofuel support policies would reduce greenhouse gas emissions from transport fuel by no more than 0.8 percent by 2015, while Brazilian ethanol from sugar cane reduces greenhouse gas emissions by at least 80 percent compared to fossil fuels. The assessment calls on governments for more open markets in biofuels and feedstocks in order to improve efficiency and lower costs. The OECD study concluded that "...current biofuel support measures alone are estimated to increase average wheat prices by about 5 percent, maize by around 7 percent and vegetable oil by about 19 percent over the next 10 years."

Another World Bank research report published on July 2010 found their previous study may have overestimated the contribution of biofuel production, as the paper concluded that "the effect of biofuels on food prices has not been as large as originally thought, but that the use of commodities by financial investors (the so-called "financialization of commodities") may have been partly responsible for the 2007/08 spike."

Sustainable Biofuel

Biofuels, in the form of liquid fuels derived from plant materials, are entering the market, driven by factors such as oil price spikes and the need for increased energy security. However, many of the biofuels that are currently being supplied have been criticised for their adverse impacts on the natural environment, food security, and land use.

The challenge is to support biofuel development, including the development of new cellulosic technologies, with responsible policies and economic instruments to help ensure that biofuel commercialization is sustainable. Responsible commercialization of biofuels represents an opportunity to enhance sustainable economic prospects in Africa, Latin America and Asia.

Biofuels have a limited ability to replace fossil fuels and should not be regarded as a 'silver bullet' to deal with transport emissions. However, they offer the prospect of increased market competition and oil price moderation. A healthy supply of alternative energy sources will help to combat gasoline price spikes and reduce dependency on fossil fuels, especially in the transport sector. Using transportation fuels more efficiently is also an integral part of a sustainable transport strategy.

Biofuel Options

Biofuel development and use is a complex issue because there are many biofuel options which are available. Biofuels, such as ethanol and biodiesel, are currently produced from the products of conventional food crops such as the starch, sugar and oil feedstocks from crops that include wheat, maize, sugar cane, palm oil and oilseed rape. Some researchers fear that a major switch to biofuels from such crops would create a direct competition with their use for food and animal feed, and claim that in some parts of the world the economic consequences are already visible, other researchers look at the land available and the enormous areas of idle and abandoned land and claim that there is room for a large proportion of biofuel also from conventional crops.

Second generation biofuels are now being produced from a much broader range of feedstocks including the cellulose in dedicated energy crops (perennial grasses such as switchgrass and Miscanthus giganteus), forestry materials, the co-products from food production, and domestic vegetable waste. Advances in the conversion processes will improve the sustainability of biofuels, through better efficiencies and reduced environmental impact of producing biofuels, from both existing food crops and from cellulosic sources.

In 2007, Ronald Oxburgh suggested in *The Courier-Mail* that production of biofuels could be either responsible or irresponsible and had several trade-offs: "Produced responsibly they are a sustainable energy source that need not divert any land from growing food nor damage the environment; they can also help solve the problems of the waste generated by Western society; and they can create jobs for the poor where previously were none. Produced irresponsibly, they at best offer no climate benefit and, at worst, have detrimental social and environmental consequences. In other words, biofuels are pretty much like any other product. In 2008 the Nobel prize-winning chemist Paul J. Crutzen published findings that the release of nitrous oxide (N_2O) emissions in the production of biofuels means that they contribute more to global warming than the fossil fuels they replace.

According to the Rocky Mountain Institute, sound biofuel production practices would not hamper food and fibre production, nor cause water or environmental problems, and would enhance soil fertility. The selection of land on which to grow the feedstocks is a critical component of the ability of biofuels to deliver sustainable solutions. A key consideration is the minimisation of biofuel competition for prime cropland.

Plants Used as Sustainable Biofuel

Sugarcane in Brazil

Sugarcane (*Saccharum officinarum*) plantation ready for harvest, Ituverava, São Paulo State, Brazil.

Mechanized harvesting of sugarcane, Piracicaba, São Paulo, Brazil.

Cosan's Costa Pinto sugar cane mill and ethanol distillery plant at Piracicaba, São Paulo, Brazil.

Brazil's production of ethanol fuel from sugarcane dates back to the 1970s, as a governmental response to the 1973 oil crisis. Brazil is considered the biofuel industry leader and the world's first sustainable biofuels economy. In 2010 the U.S. Environmental Protection Agency designated Brazilian sugarcane ethanol as an advanced biofuel due to EPA's estimated 61% reduction of total life cycle greenhouse gas emissions, including direct indirect land use change emissions. Brazil sugarcane ethanol fuel program success and sustainability is based on the most efficient agricultural technology for sugarcane cultivation in the world, uses modern equipment and cheap sugar cane as feedstock, the residual cane-waste (bagasse) is used to process heat and power, which results in a very competitive price and also in a high energy balance (output energy/input energy), which varies from 8.3 for average conditions to 10.2 for best practice production.

A report commissioned by the United Nations, based on a detailed review of published research up to mid-2009 as well as the input of independent experts world-wide, found that ethanol from sugar cane as produced in Brazil *"in some circumstances does better than just "zero emission". If grown and processed correctly, it has negative emission, pulling CO2 out of the atmosphere, rather than adding it.* In contrast, the report found that U.S. use of maize for biofuel is less efficient, as sugarcane can lead to emissions reductions of between 70% and well over 100% when substituted for gasoline. Several other studies have shown that sugarcane-based ethanol reduces greenhouse gases by 86 to 90% if there is no significant land use change.

In another study commissioned by the Dutch government in 2006 to evaluate the sustainability of Brazilian bioethanol concluded that there is sufficient water to supply all foreseeable long-term water requirements for sugarcane and ethanol production. This evaluation also found that consumption of agrochemicals for sugar cane production is lower than in citric, corn, coffee and soybean cropping. The study found that development of resistant sugar cane varieties is a crucial aspect of disease and pest control and is one of the primary objectives of Brazil's cane genetic improvement programs. Disease control is one of the main reasons for the replacement of a commercial variety of sugar cane.

Another concern is the fact that sugarcane fields are traditionally burned just before harvest to avoid harm to the workers, by removing the sharp leaves and killing snakes and other harmful animals, and also to fertilize the fields with ash. Mechanization will reduce pollution from burning fields and has higher productivity than people, and due to mechanization the number of temporary workers in the sugarcane plantations has already declined. By the 2008 harvest season, around 47% of the cane was collected with harvesting machines.

Regarding the negative impacts of the potential direct and indirect effect of land use changes on carbon emissions, the study commissioned by the Dutch government concluded that "it is very difficult to determine the indirect effects of further land use for sugar cane production (i.e. sugar cane replacing another crop like soy or citrus crops, which in turn causes additional soy plantations replacing pastures, which in turn may cause deforestation), and also not logical to attribute all these soil carbon losses to sugar cane". The Brazilian agency Embrapa estimates that there is enough agricultural land available to increase at least 30 times the existing sugarcane plantation without endangering sensible ecosystems or taking land destined for food crops. Most future growth is expected to take place on abandoned pasture lands, as it has been the historical trend in São Paulo state. Also, productivity is expected to improve even further based on current biotechnology research, genetic improvement, and better agronomic practices, thus contributing to reduce land demand for future sugarcane cultures.

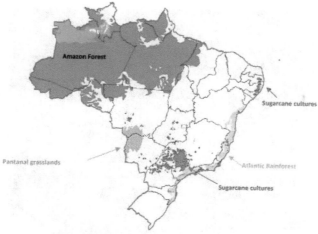

Location of environmentally valuable areas with respect to sugarcane plantations. São Paulo, located in the Southeast Region of Brazil, concentrates two-thirds of sugarcane cultures.

Another concern is the risk of clearing rain forests and other environmentally valuable land for sugarcane production, such as the Amazonia, the Pantanal or the Cerrado. Embrapa has rebutted this concern explaining that 99.7% of sugarcane plantations are located at least 2,000 km from the Amazonia, and expansion during the last 25 years took place in the Center-South region, also far away from the Amazonia, the Pantanal or the Atlantic forest. In São Paulo state growth took place in abandoned pasture lands. The impact assessment commissioned by the Dutch government supported this argument.

In order to guarantee a sustainable development of ethanol production, in September 2009 the government issued by decree a countrywide agroecological land use zoning to restrict sugarcane growth in or near environmentally sensitive areas. According to the new criteria, 92.5% of the Brazilian territory is not suitable for sugarcane plantation. The government considers that the suitable areas are more than enough to meet the future demand for ethanol and sugar in the domestic and international markets foreseen for the next decades.

Regarding the food vs fuel issue, a World Bank research report published on July 2008 found that *"Brazil's sugar-based ethanol did not push food prices appreciably higher"*. This research paper also concluded that Brazil's sugar cane–based ethanol has not raised sugar prices significantly. An economic assessment report also published in July 2008 by the OECD agrees with the World Bank report regarding the negative effects of subsidies and trade restrictions, but found that the impact of biofuels on food prices are much smaller. A study by the Brazilian research unit of the Fundação Getúlio Vargas regarding the effects of biofuels on grain prices concluded that the major driver behind the 2007-2008 rise in food prices was speculative activity on futures markets under conditions of increased demand in a market with low grain stocks. The study also concluded that there is no correlation between Brazilian sugarcane cultivated area and average grain prices, as on the contrary, the spread of sugarcane was accompanied by rapid growth of grain crops in the country.

Jatropha

India and Africa

Jatropha gossipifolia in Hyderabad, India.

Crops like Jatropha, used for biodiesel, can thrive on marginal agricultural land where many trees and crops won't grow, or would produce only slow growth yields. Jatropha cultivation provides benefits for local communities:

Cultivation and fruit picking by hand is labour-intensive and needs around one person per hectare. In parts of rural India and Africa this provides much-needed jobs - about 200,000 people worldwide now find employment through jatropha. Moreover, villagers often find that they can grow other crops in the shade of the trees. Their communities will avoid importing expensive diesel and there will be some for export too.

Cambodia

Cambodia has no proven fossil fuel reserves, and is almost completely dependent on imported diesel fuel for electricity production. Consequently, Cambodians face an insecure supply and pay some of the highest energy prices in the world. The impacts of this are widespread and may hinder economic development.

Biofuels may provide a substitute for diesel fuel that can be manufactured locally for a lower price, independent of the international oil price. The local production and use of biofuel also offers other benefits such as improved energy security, rural development opportunities and environmental benefits. The Jatropha curcas species appears to be a particularly suitable source of biofuel as it already grows commonly in Cambodia. Local sustainable production of biofuel in Cambodia, based on the Jatropha or other sources, offers good potential benefits for the investors, the economy, rural communities and the environment.

Mexico

Jatropha is native to Mexico and Central America and was likely transported to India and Africa in the 1500s by Portuguese sailors convinced it had medicinal uses. In 2008, recognizing the need to diversify its sources of energy and reduce emissions, Mexico passed a law to push developing biofuels that don't threaten food security and the agriculture ministry has since identified some 2.6 million hectares (6.4 million acres) of land with a high potential to produce jatropha. The Yucatán Peninsula, for instance, in addition to being a corn producing region, also contains abandoned sisal plantations, where the growing of Jatropha for biodiesel production would not displace food.

On April 1, 2011 Interjet completed the first Mexican aviation biofuels test flight on an Airbus A320. The fuel was a 70:30 traditional jet fuel biojet blend produced from Jatropha oil provided by three Mexican producers, Global Energías Renovables (a wholly owned subsidiary of U.S.-based Global Clean Energy Holdings, Bencafser S.A. and Energy JH S.A. Honeywell's UOP processed the oil into Bio-SPK (Synthetic Paraffinic Kerosene) . Global Energías Renovables operates the largest Jatropha farm in the Americas.

On August 1, 2011 Aeromexico, Boeing, and the Mexican Government participated in the first bio-jet powered transcontinental flight in aviation history. The flight from Mexico City to Madrid used a blend of 70 percent traditional fuel and 30 percent biofuel (aviation biofuel). The biojet was produced entirely from Jatropha oil.

Pongamia Pinnata in Australia and India

Pongamia pinnata seeds in Brisbane, Australia.

Pongamia pinnata is a legume native to Australia, India, Florida (USA) and most tropical regions, and is now being invested in as an alternative to Jatropha for areas such as Northern Australia, where Jatropha is classed as a noxious weed. Commonly known as simply 'Pongamia', this tree is currently being commercialised in Australia by Pacific Renewable Energy, for use as a Diesel replacement for running in modified Diesel engines or for conversion to Biodiesel using 1st or 2nd Generation Biodiesel techniques, for running in unmodified Diesel engines.

Sweet Sorghum in India

Sweet sorghum overcomes many of the shortcomings of other biofuel crops. With sweet sorghum, only the stalks are used for biofuel production, while the grain is saved for food or livestock feed. It is not in high demand in the global food market, and thus has little impact on food prices and food security. Sweet sorghum is grown on already-farmed drylands that are low in carbon storage capacity, so concerns about the clearing of rainforest do not apply. Sweet sorghum is easier and cheaper to grow than other biofuel crops in India and does not require irrigation, an important consideration in dry areas. Some of the Indian sweet sorghum varieties are now grown in Uganda for ethanol production.

A study by researchers at the International Crops Research Institute for the Semi-Arid Tropics (ICRISAT) found that growing sweet sorghum instead of grain sorghum could increase farmers incomes by US$40 per hectare per crop because it can provide food, feed and fuel. With grain sorghum currently grown on over 11 million hectares (ha) in Asia and on 23.4 million ha in Africa, a switch to sweet sorghum could have a considerable economic impact.

International Collaboration on Sustainable Biofuels

Roundtable on Sustainable Biofuels

Public attitudes and the actions of key stakeholders can play a crucial role in realising the potential of sustainable biofuels. Informed discussion and dialogue, based both on scientific research and an understanding of public and stakeholder views, is important.

The Roundtable on Sustainable Biofuels is an international initiative which brings together farmers, companies, governments, non-governmental organizations, and scientists who are interested in the sustainability of biofuels production and distribution. During 2008, the Roundtable used meetings, teleconferences, and online discussions to develop a series of principles and criteria for sustainable biofuels production.

In 2008, the Roundtable for Sustainable Biofuels released its proposed standards for sustainable biofuels. This includes 12 principles:

1. "Biofuel production shall follow international treaties and national laws regarding such things as air quality, water resources, agricultural practices, labor conditions, and more.

2. Biofuels projects shall be designed and operated in participatory processes that involve all relevant stakeholders in planning and monitoring.

3. Biofuels shall significantly reduce greenhouse gas emissions as compared to fossil fuels. The principle seeks to establish a standard methodology for comparing greenhouse gases (GHG) benefits.

4. Biofuel production shall not violate human rights or labor rights, and shall ensure decent work and the well-being of workers.

5. Biofuel production shall contribute to the social and economic development of local, rural and indigenous peoples and communities.

6. Biofuel production shall not impair food security.

7. Biofuel production shall avoid negative impacts on biodiversity, ecosystems and areas of high conservation value.

8. Biofuel production shall promote practices that improve soil health and minimize degradation.

9. Surface and groundwater use will be optimized and contamination or depletion of water resources minimized.

10. Air pollution shall be minimized along the supply chain.

11. Biofuels shall be produced in the most cost-effective way, with a commitment to improve production efficiency and social and environmental performance in all stages of the biofuel value chain.

12. Biofuel production shall not violate land rights".

In April 2011, the Roundtable on Sustainable Biofuels launched a set of comprehensive sustainability criteria - the "RSB Certification System." Biofuels producers that meet to these criteria are able to show buyers and regulators that their product has been obtained without harming the environment or violating human rights.

Sustainable Biofuels Consensus

The Sustainable Biofuels Consensus is an international initiative which calls upon governments, the private sector, and other stakeholders to take decisive action to ensure the sustainable trade, production, and use of biofuels. In this way biofuels may play a key role in energy sector transformation, climate stabilization, and resulting worldwide revitalisation of rural areas.

The Sustainable Biofuels Consensus envisions a "landscape that provides food, fodder, fiber, and energy, which offers opportunities for rural development; that diversifies energy supply, restores ecosystems, protects biodiversity, and sequesters carbon".

Better Sugarcane Initiative / Bonsucro

In 2008, a multi-stakeholder process was initiated by the World Wildlife Fund and the International Finance Corporation, the private development arm of the World Bank, bringing together industry, supply chain intermediaries, end-users, farmers and civil society organisations to develop standards for certifying the derivative products of sugar cane, one of which is ethanol fuel.

The Bonsucro standard is based around a definition of sustainability which is founded on five principles:

1. Obey the law

2. Respect human rights and labour standards

3. Manage input, production and processing efficiencies to enhance sustainability

4. Actively manage biodiversity and ecosystem services

5. Continuously improve key areas of the business

Biofuel producers that wish to sell products marked with the Bonsucro standard must both ensure that they product to the Production Standard, and that their downstream buyers meet the Chain of Custody Standard. In addition, if they wish to sell to the European market and count against the EU Renewable Energy Directive, then they must adhere to the Bonsucro EU standard, which includes specific greenhouse gas calculations following European Commission calculation guidelines.

Sustainability Standards

Several countries and regions have introduced policies or adopted standards to promote sustainable biofuels production and use, most prominently the European Union and the United States. The 2009 EU Renewable Energy Directive, which requires 10 percent of transportation energy from renewable energy by 2020, is the most comprehensive mandatory sustainability standard in place as of 2010. The Directive requires that the lifecycle greenhouse gas emissions of biofuels consumed be at least 50 percent less than the equivalent emissions from gasoline or diesel by 2017 (and 35 percent less starting in 2011). Also, the feedstocks for biofuels "should not be harvested from lands with high biodiversity value, from carbon-rich or forested land, or from wetlands".

As with the EU, the U.S. Renewable Fuel Standard (RFS) and the California Low Carbon Fuel Standard (LCFS) both require specific levels of lifecycle greenhouse gas reductions compared to equivalent fossil fuel consumption. The RFS requires that at least half of the biofuels production mandated by 2022 should reduce lifecycle emissions by 50 percent. The LCFS is a performance standard that calls for a minimum of 10 percent emissions reduction per unit of transport energy by 2020. Both the U.S. and California standards currently address only greenhouse gas emissions, but California plans to "expand its policy to address other sustainability issues associated with liquid biofuels in the future".

In 2009, Brazil also adopted new sustainability policies for sugarcane ethanol, including "zoning regulation of sugarcane expansion and social protocols".

Oil Price Moderation

Biofuels offer the prospect of real market competition and oil price moderation. According to the Wall Street Journal, crude oil would be trading 15 per cent higher and gasoline would be as much as 25 per cent more expensive, if it were not for biofuels. A healthy supply of alternative energy sources will help to combat gasoline price spikes.

Sustainable Transport

Biofuels have a limited ability to replace fossil fuels and should not be regarded as a 'silver bullet' to deal with transport emissions. Biofuels on their own cannot deliver a sustainable transport system and so must be developed as part of an integrated approach, which promotes other renewable energy options and energy efficiency, as well as reducing the overall energy demand and need for transport. Consideration needs to be given to the development of hybrid and fuel cell vehicles, public transport, and better town and rural planning.

In December 2008 an Air New Zealand jet completed the world's first commercial aviation test flight partially using jatropha-based fuel. More than a dozen performance tests were undertaken in the two-hour test flight which departed from Auckland International Airport. A biofuel blend of 50:50 jatropha and Jet A1 fuel was used to power one of the Boeing 747-400's Rolls-Royce RB211 engines. Air New Zealand set several criteria for its jatropha, requiring that "the land it came from was neither forest nor virgin grassland in the previous 20 years, that the soil and climate it came from is not suitable for the majority of food crops and that the farms are rain fed and not mechanically irrigated". The company has also set general sustainability criteria, saying that such biofuels must not compete with food resources, that they must be as good as traditional jet fuels, and that they should be cost competitive.

In January 2009, Continental Airlines used a sustainable biofuel to power a commercial aircraft for the first time in North America. This demonstration flight marks the first sustainable biofuel demonstration flight by a commercial carrier using a twin-engined aircraft, a Boeing 737-800, powered by CFM International CFM56-7B engines. The biofuel blend included components derived from algae and jatropha plants. The algae oil was provided by Sapphire Energy, and the jatropha oil by Terasol Energy.

In March 2011, Yale University research showed significant potential for sustainable aviation fuel based on jatropha-curcas. According to the research, if cultivated properly, "jatropha can deliver many benefits in Latin America and greenhouse gas reductions of up to 60 percent when compared to petroleum-based jet fuel". Actual farming conditions in Latin America were assessed using sustainability criteria developed by the Roundtable on Sustainable Biofuels. Unlike previous research, which used theoretical inputs, the Yale team conducted many interviews with jatropha farmers and used "field measurements to develop the first comprehensive sustainability analysis of actual projects".

As of June 2011, revised international aviation fuel standards officially allow commercial airlines to blend conventional jet fuel with up to 50 percent biofuels. The renewable fuels "can be blended with conventional commercial and military jet fuel through requirements in the newly issued edition of ASTM D7566, Specification for Aviation Turbine Fuel Containing Synthesized Hydrocarbons".

In December 2011, the FAA awarded $7.7 million to eight companies to advance the development of commercial aviation biofuels, with a special focus on alcohol to jet fuel. The FAA is assisting in the development of a sustainable fuel (from alcohols, sugars, biomass, and organic matter such as pyrolysis oils) that can be "dropped in" to aircraft without changing current practices and infrastructure. The research will test how the new fuels affect engine durability and quality control standards.

GreenSky London, a biofuels plant under construction in 2014, will take in some 500,000 tonnes of municipal rubbish and change the organic component into 60,000 tonnes of jet fuel, and 40 megawatts of power. By the end of 2015, all British Airways flights from London City Airport will be fuelled by waste and rubbish discarded by London residents.

Environmental Impact of Biodiesel

The environmental impact of biodiesel is diverse.

Greenhouse Gas Emissions

An often mentioned incentive for using biodiesel is its capacity to lower greenhouse gas emissions compared to those of fossil fuels. Whether this is true or not depends on many factors. Especially the effects from land use change have potential to cause even more emissions than what would be caused by using fossil fuels alone.

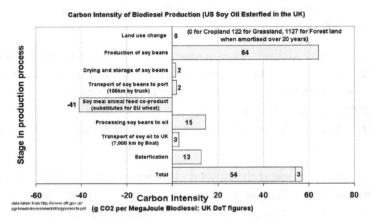

Calculation of Carbon Intensity of Soy biodiesel grown in the US and burnt in the UK, using figures calculated by the UK government for the purposes of the Renewable transport fuel obligation.

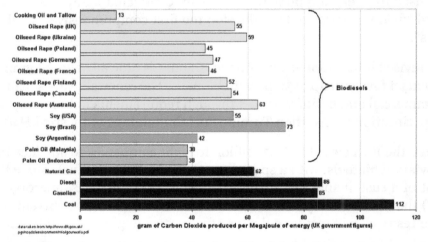

Graph of UK figures for the Carbon Intensity of Biodiesels and fossil fuels. This graph assumes that all biodiesel is used in its country of origin. It also assumes that the diesel is produced from pre-existing croplands rather than by changing land use

Carbon dioxide is one of the major greenhouse gases. Although the burning of biodiesel produces carbon dioxide emissions similar to those from ordinary fossil fuels, the plant feedstock used in the production absorbs carbon dioxide from the atmosphere when it grows. Plants absorb carbon dioxide through a process known as photosynthesis which allows it to store energy from sunlight in the form of sugars and starches. After the biomass is converted into biodiesel and burned as fuel the energy and carbon is released again. Some of that energy can be used to power an engine while the carbon dioxide is released back into the atmosphere.

When considering the total amount of greenhouse gas emissions it is therefore important to consider the whole production process and what indirect effects such production might cause. The effect on carbon dioxide emissions is highly dependent on production methods and the type of feedstock used. Calculating the carbon intensity of biofuels is a complex and inexact process, and is highly dependent on the assumptions made in the calculation. A calculation usually includes:

- Emissions from growing the feedstock (e.g. Petrochemicals used in fertilizers)

- Emissions from transporting the feedstock to the factory

- Emissions from processing the feedstock into biodiesel

Other factors can be very significant but are sometimes not considered. These include:

- Emissions from the change in land use of the area where the fuel feedstock is grown.

- Emissions from transportation of the biodiesel from the factory to its point of use

- The efficiency of the biodiesel compared with standard diesel

- The amount of Carbon Dioxide produced at the tail pipe. (Biodiesel can produce 4.7% more)

- The benefits due to the production of useful by-products, such as cattle feed or glycerine

If land use change is not considered and assuming today's production methods, biodiesel from rapeseed and sunflower oil produce 45%-65% lower greenhouse gas emissions than petrodiesel. However, there is ongoing research to improve the efficiency of the production process. Biodiesel produced from used cooking oil or other waste fat could reduce CO_2 emissions by as much as 85%. As long as the feedstock is grown on existing cropland, land use change has little or no effect on greenhouse gas emissions. However, there is concern that increased feedstock production directly affects the rate of deforestation. Such clearcutting cause carbon stored in the forest, soil and peat layers to be released. The amount of greenhouse gas emissions from deforestation is so large that the benefits from lower emissions (caused by biodiesel use alone) would be negligible for hundreds of years. Biofuel produced from feedstock such as palm oil could therefore cause much higher carbon dioxide emissions than some types of fossil fuels.

Pollution

In the United States, biodiesel is the only alternative fuel to have successfully completed the Health Effects Testing requirements (Tier I and Tier II) of the Clean Air Act (1990).

Biodiesel can reduce the direct tailpipe-emission of particulates, small particles of solid combustion products, on vehicles with particulate filters by as much as 20 percent compared with low-sulfur (< 50 ppm) diesel. Particulate emissions as the result of production are reduced by around 50 percent compared with fossil-sourced diesel. (Beer et al., 2004). Biodiesel has a higher cetane rating than petrodiesel, which can improve performance and clean up emissions compared to crude petro-diesel (with cetane lower than 40). Biodiesel contains fewer aromatic hydrocarbons: benzofluoranthene: 56% reduction; Benzopyrenes: 71% reduction.

Biodegradation

A University of Idaho study compared biodegradation rates of biodiesel, neat vegetable oils, biodiesel and petroleum diesel blends, and neat 2-D diesel fuel. Using low concentrations of the product to be degraded (10 ppm) in nutrient and sewage sludge amended solutions, they demonstrated that biodiesel degraded at the same rate as a dextrose control and 5 times as quickly as petroleum diesel over a period of 28 days, and that biodiesel blends doubled the rate of petroleum diesel degradation through co-metabolism. The same study examined soil degradation using 10 000 ppm of biodiesel and petroleum diesel, and found biodiesel degraded at twice the rate of petroleum diesel in soil. In all cases, it was determined biodiesel also degraded more completely than petroleum diesel, which produced poorly degradable undetermined intermediates. Toxicity studies for the same project demonstrated no mortalities and few toxic effects on rats and rabbits with up to 5000 mg/kg of biodiesel. Petroleum diesel showed no mortalities at the same concentration either, however toxic effects such as hair loss and urinary discolouring were noted with concentrations of >2000 mg/l in rabbits.:

Biodegradation in Aquatic Environments

As biodiesel becomes more widely used, it is important to consider how consumption affects water quality and aquatic ecosystems. Research examining the biodegradability of different biodiesel fuels found that all of the biofuels studied (including Neat Rapeseed oil, Neat Soybean oil, and their modified ester products) were "readily biodegradable" compounds, and had a relatively high biodegradation rate in water. Additionally, the presence of biodiesel can increase the rate of diesel biodegradation via co-metabolism. As the ratio of biodiesel is increased in biodiesel/diesel mixtures, the faster the diesel is degraded. Another study using controlled experimental conditions also showed that fatty acid methyl esters, the primary molecules in biodiesel, degraded much faster than petroleum diesel in sea water.

Carbonyl Emissions

When considering the emissions from fossil fuel and biofuel use, research typically focuses on major pollutants such as hydrocarbons. It is generally recognized that using biodiesel in place of diesel results in a substantial reduction in regulated gas emissions, but there has been a lack of information in research literature about the non-regulated compounds which also play a role in air pollution. One study focused on the emissions of non-criteria carbonyl compounds from the burning of pure diesel and biodiesel blends in heavy-duty diesel engines. The results found that carbonyl emissions of formaldehyde, acetaldehyde, acrolein, acetone, propionaldehyde and butyraldehyde, were higher in biodiesel mixtures than emissions from pure diesel. Biodiesel use results in higher carbonyl emissions but lower total hydrocarbon emissions, which may be better as an alternative fuel source. Other studies have been done which conflict with these results, but comparisons are difficult to make due to various factors that differ between studies (such as types of fuel and engines used). In a paper which compared 12 research articles on carbonyl emissions from biodiesel fuel use, it found that 8 of the papers reported increased carbonyl compound emissions while 4 showed the opposite. This is evidence that there is still much research required on these compounds.

Deforestation

Deforestation, clearance or clearing is the removal of a forest or stand of trees where the land is thereafter converted to a non-forest use. Examples of deforestation include conversion of forestland to farms, ranches, or urban use. Tropical rainforests is where the most concentrated deforestation occurs. About 30% of Earth's land surface is covered by forests.

Satellite photograph of deforestation in progress in the Tierras Bajas project in eastern Bolivia.

In temperate mesic climates, natural regeneration of forest stands often will not occur in the absence of disturbance, whether natural or anthropogenic. Furthermore, biodiversity after regeneration harvest often mimics that found after natural disturbance, including biodiversity loss after naturally occurring rainforest destruction.

Deforestation occurs for multiple reasons: trees are cut down to be used or sold as fuel (sometimes in the form of charcoal) or timber, while cleared land is used as pasture for livestock and plantation. The removal of trees without sufficient reforestation has resulted in damage to habitat, biodiversity loss and aridity. It has adverse impacts on biosequestration of atmospheric carbon dioxide. Deforestation has also been used in war to deprive the enemy of cover for its forces and also vital resources. Modern examples of this were the use of Agent Orange by the British military in Malaya during the Malayan Emergency and the United States military in Vietnam during the Vietnam War. As of 2005, net deforestation rates have ceased to increase in countries with a per capita GDP of at least US$4,600. Deforested regions typically incur significant adverse soil erosion and frequently degrade into wasteland.

Disregard of ascribed value, lax forest management and deficient environmental laws are some of the factors that allow deforestation to occur on a large scale. In many countries, deforestation, both naturally occurring and human induced, is an ongoing issue. Deforestation causes extinction, changes to climatic conditions, desertification, and displacement of populations as observed by current conditions and in the past through the fossil record. More than half of all plant and land animal species in the world live in tropical forests.

Between 2000 and 2012, 2.3 million square kilometres (890,000 square miles) of forests around the earth were cut down. As a result of deforestation, only 6.2 million square kilometres (2.4 million square miles) remain of the original 16 million square kilometres (6 million square miles) of forest that formerly covered the earth.

Causes

According to the United Nations Framework Convention on Climate Change (UNFCCC) secretariat, the overwhelming direct cause of deforestation is agriculture. Subsistence farming is responsible for 48% of deforestation; commercial agriculture is responsible for 32% of deforestation; logging is responsible for 14% of deforestation and fuel wood removals make up 5% of deforestation.

Experts do not agree on whether industrial logging is an important contributor to global deforestation. Some argue that poor people are more likely to clear forest because they have no alternatives, others that the poor lack the ability to pay for the materials and labour needed to clear forest. One study found that population increases due to high fertility rates were a primary driver of tropical deforestation in only 8% of cases.

Other causes of contemporary deforestation may include corruption of government institutions, the inequitable distribution of wealth and power, population growth and overpopulation, and urbanization. Globalization is often viewed as another root cause of deforestation, though there are cases in which the impacts of globalization (new flows of labor, capital, commodities, and ideas) have promoted localized forest recovery.

The last batch of sawnwood from the peat forest in Indragiri Hulu, Sumatra, Indonesia.
Deforestation for oil palm plantation.

In 2000 the United Nations Food and Agriculture Organization (FAO) found that "the role of population dynamics in a local setting may vary from decisive to negligible," and that deforestation can result from "a combination of population pressure and stagnating economic, social and technological conditions."

The degradation of forest ecosystems has also been traced to economic incentives that make forest conversion appear more profitable than forest conservation. Many important forest functions have no markets, and hence, no economic value that is readily apparent to the forests' owners or the communities that rely on forests for their well-being. From the perspective of the developing world, the benefits of forest as carbon sinks or biodiversity reserves go primarily to richer developed nations and there is insufficient compensation for these services. Developing countries feel that some countries in the developed world, such as the United States of America, cut down their forests centuries ago and benefited greatly from this deforestation, and that it is hypocritical to deny developing countries the same opportunities: that the poor shouldn't have to bear the cost of preservation when the rich created the problem.

Some commentators have noted a shift in the drivers of deforestation over the past 30 years. Whereas deforestation was primarily driven by subsistence activities and government-sponsored development projects like transmigration in countries like Indonesia and colonization in Latin

America, India, Java, and so on, during late 19th century and the earlier half of the 20th century. By the 1990s the majority of deforestation was caused by industrial factors, including extractive industries, large-scale cattle ranching, and extensive agriculture.

Environmental Problems

Atmospheric

Illegal slash and burn practice in Madagascar, 2010

Deforestation is ongoing and is shaping climate and geography.

Deforestation is a contributor to global warming, and is often cited as one of the major causes of the enhanced greenhouse effect. Tropical deforestation is responsible for approximately 20% of world greenhouse gas emissions. According to the Intergovernmental Panel on Climate Change deforestation, mainly in tropical areas, could account for up to one-third of total anthropogenic carbon dioxide emissions. But recent calculations suggest that carbon dioxide emissions from deforestation and forest degradation (excluding peatland emissions) contribute about 12% of total anthropogenic carbon dioxide emissions with a range from 6 to 17%. Deforestation causes carbon dioxide to linger in the atmosphere. As carbon dioxide accrues, it produces a layer in the atmosphere that traps radiation from the sun. The radiation converts to heat which causes global warming, which is better known as the greenhouse effect. Plants remove carbon in the form of carbon dioxide from the atmosphere during the process of photosynthesis, but release some carbon dioxide back into the atmosphere during normal respiration. Only when actively growing can a tree or forest remove carbon, by storing it in plant tissues. Both the decay and burning of wood releases much of this stored carbon back to the atmosphere. In order for forests to take up carbon, there must be a net accumulation of wood. One way is for the wood to be harvested and turned into long-lived products, with new young trees replacing them. Deforestation may also cause carbon stores held in soil to be released. Forests can be either sinks or sources depending upon environmental circumstances. Mature forests alternate between being net sinks and net sources of carbon dioxide

In deforested areas, the land heats up faster and reaches a higher temperature, leading to localized upward motions that enhance the formation of clouds and ultimately produce more rainfall. However, according to the Geophysical Fluid Dynamics Laboratory, the models used to investigate remote responses to tropical deforestation showed a broad but mild temperature increase all through the tropical atmosphere. The model predicted <0.2 °C warming for upper air at 700 mb and 500 mb. However, the model shows no significant changes in other areas besides the Tropics. Though the model showed no significant changes to the climate in areas other than the Tropics, this may not be the case since the model has possible errors and the results are never absolutely definite.

Fires on Borneo and Sumatra, 2006. People use slash-and-burn deforestation to clear land for agriculture.

Reducing emissions from deforestation and forest degradation (REDD) in developing countries has emerged as a new potential to complement ongoing climate policies. The idea consists in providing financial compensations for the reduction of greenhouse gas (GHG) emissions from deforestation and forest degradation".

Rainforests are widely believed by laymen to contribute a significant amount of the world's oxygen, although it is now accepted by scientists that rainforests contribute little net oxygen to the atmosphere and deforestation has only a minor effect on atmospheric oxygen levels. However, the incineration and burning of forest plants to clear land releases large amounts of CO_2, which contributes to global warming. Scientists also state that tropical deforestation releases 1.5 billion tons of carbon each year into the atmosphere.

Hydrological

The water cycle is also affected by deforestation. Trees extract groundwater through their roots and release it into the atmosphere. When part of a forest is removed, the trees no longer transpire this water, resulting in a much drier climate. Deforestation reduces the content of water in the soil and groundwater as well as atmospheric moisture. The dry soil leads to lower water intake for the trees to extract. Deforestation reduces soil cohesion, so that erosion, flooding and landslides ensue.

Shrinking forest cover lessens the landscape's capacity to intercept, retain and transpire precipitation. Instead of trapping precipitation, which then percolates to groundwater systems, deforested areas become sources of surface water runoff, which moves much faster than subsurface flows. That quicker transport of surface water can translate into flash flooding and more localized floods than would occur with the forest cover. Deforestation also contributes to decreased evapotranspiration, which lessens atmospheric moisture which in some cases affects precipitation levels downwind from the deforested area, as water is not recycled to downwind forests, but is lost in runoff and returns directly to the oceans. According to one study, in deforested north and northwest China, the average annual precipitation decreased by one third between the 1950s and the 1980s.

Trees, and plants in general, affect the water cycle significantly:

- their canopies intercept a proportion of precipitation, which is then evaporated back to the atmosphere (canopy interception);

- their litter, stems and trunks slow down surface runoff;

- their roots create macropores – large conduits – in the soil that increase infiltration of water;

- they contribute to terrestrial evaporation and reduce soil moisture via transpiration;

- their litter and other organic residue change soil properties that affect the capacity of soil to store water.

- their leaves control the humidity of the atmosphere by transpiring. 99% of the water absorbed by the roots moves up to the leaves and is transpired.

As a result, the presence or absence of trees can change the quantity of water on the surface, in the soil or groundwater, or in the atmosphere. This in turn changes erosion rates and the availability of water for either ecosystem functions or human services.

The forest may have little impact on flooding in the case of large rainfall events, which overwhelm the storage capacity of forest soil if the soils are at or close to saturation.

Tropical rainforests produce about 30% of our planet's fresh water.

Soil

Deforestation for the use of clay in the Brazilian city of Rio de Janeiro.
The hill depicted is Morro da Covanca, in Jacarepaguá

Undisturbed forests have a very low rate of soil loss (erosion), approximately 2 metric tons per square kilometer (6 short tons per square mile). Deforestation generally increases rates of soil loss, by increasing the amount of runoff and reducing the protection of the soil from tree litter. This can be an advantage in excessively leached tropical rain forest soils. Forestry operations themselves also increase erosion through the development of (forest) roads and the use of mechanized equipment.

China's Loess Plateau was cleared of forest millennia ago. Since then it has been eroding, creating dramatic incised valleys, and providing the sediment that gives the Yellow River its yellow color and that causes the flooding of the river in the lower reaches (hence the river's nickname 'China's sorrow').

Removal of trees does not always increase erosion rates. In certain regions of southwest US, shrubs and trees have been encroaching on grassland. The trees themselves enhance the loss of grass between tree canopies. The bare intercanopy areas become highly erodible. The US Forest Service, in Bandelier National Monument for example, is studying how to restore the former ecosystem, and reduce erosion, by removing the trees.

Tree roots bind soil together, and if the soil is sufficiently shallow they act to keep the soil in place by also binding with underlying bedrock. Tree removal on steep slopes with shallow soil thus increases the risk of landslides, which can threaten people living nearby.

Biodiversity

Deforestation on a human scale results in decline in biodiversity, and on a natural global scale is known to cause the extinction of many species. The removal or destruction of areas of forest cover has resulted in a degraded environment with reduced biodiversity. Forests support biodiversity, providing habitat for wildlife; moreover, forests foster medicinal conservation. With forest biotopes being irreplaceable source of new drugs (such as taxol), deforestation can destroy genetic variations (such as crop resistance) irretrievably.

Illegal logging in Madagascar. In 2009, the vast majority of the illegally obtained rosewood was exported to China.

Since the tropical rainforests are the most diverse ecosystems on Earth and about 80% of the world's known biodiversity could be found in tropical rainforests, removal or destruction of significant areas of forest cover has resulted in a degraded environment with reduced biodiversity. A study in Rondônia, Brazil, has shown that deforestation also removes the microbial community which is involved in the recycling of nutrients, the production of clean water and the removal of pollutants.

It has been estimated that we are losing 137 plant, animal and insect species every single day due to rainforest deforestation, which equates to 50,000 species a year. Others state that tropical rainforest deforestation is contributing to the ongoing Holocene mass extinction. The known extinction rates from deforestation rates are very low, approximately 1 species per year from mammals and birds which extrapolates to approximately 23,000 species per year for all species. Predictions have been made that more than 40% of the animal and plant species in Southeast Asia could be wiped out in the 21st century. Such predictions were called into question by 1995 data that show that within regions of Southeast Asia much of the original forest has been converted to monospecific plantations, but that potentially endangered species are few and tree flora remains widespread and stable.

Scientific understanding of the process of extinction is insufficient to accurately make predictions about the impact of deforestation on biodiversity. Most predictions of forestry related biodiversity loss are based on species-area models, with an underlying assumption that as the forest declines species diversity will decline similarly. However, many such models have been proven to be wrong and loss of habitat does not necessarily lead to large scale loss of species. Species-area models are known to overpredict the number of species known to be threatened in areas where actual deforestation is ongoing, and greatly overpredict the number of threatened species that are widespread.

A recent study of the Brazilian Amazon predicts that despite a lack of extinctions thus far, up to 90 percent of predicted extinctions will finally occur in the next 40 years.

Economic Impact

Damage to forests and other aspects of nature could halve living standards for the world's poor and reduce global GDP by about 7% by 2050, a report concluded at the Convention on Biological Diversity (CBD) meeting in Bonn. Historically, utilization of forest products, including timber and fuel wood, has played a key role in human societies, comparable to the roles of water and cultivable land. Today, developed countries continue to utilize timber for building houses, and wood pulp for paper. In developing countries almost three billion people rely on wood for heating and cooking.

The forest products industry is a large part of the economy in both developed and developing countries. Short-term economic gains made by conversion of forest to agriculture, or over-exploitation of wood products, typically leads to loss of long-term income and long-term biological productivity. West Africa, Madagascar, Southeast Asia and many other regions have experienced lower revenue because of declining timber harvests. Illegal logging causes billions of dollars of losses to national economies annually.

The new procedures to get amounts of wood are causing more harm to the economy and overpower the amount of money spent by people employed in logging. According to a study, "in most areas studied, the various ventures that prompted deforestation rarely generated more than US$5 for every ton of carbon they released and frequently returned far less than US$1". The price on the European market for an offset tied to a one-ton reduction in carbon is 23 euro (about US$35).

Rapidly growing economies also have an effect on deforestation. Most pressure will come from the world's developing countries, which have the fastest-growing populations and most rapid economic (industrial) growth. In 1995, economic growth in developing countries reached nearly 6%, compared with the 2% growth rate for developed countries." As our human population grows, new homes, communities, and expansions of cities will occur. Connecting all of the new expansions will be roads, a very important part in our daily life. Rural roads promote economic development but also facilitate deforestation. About 90% of the deforestation has occurred within 100 km of roads in most parts of the Amazon.

The European Union is one of the largest importer of products made from illegal deforestation.

Forest Transition Theory

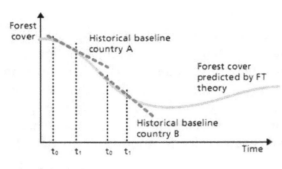

Source: Angelsen 2008.

The forest transition and historical baselines.

The forest area change may follow a pattern suggested by the forest transition (FT) theory, whereby at early stages in its development a country is characterized by high forest cover and low deforestation rates (HFLD countries).

Then deforestation rates accelerate (HFHD, high forest cover – high deforestation rate), and forest cover is reduced (LFHD, low forest cover – high deforestation rate), before the deforestation rate slows (LFLD, low forest cover – low deforestation rate), after which forest cover stabilizes and eventually starts recovering. FT is not a "law of nature," and the pattern is influenced by national context (for example, human population density, stage of development, structure of the economy), global economic forces, and government policies. A country may reach very low levels of forest cover before it stabilizes, or it might through good policies be able to "bridge" the forest transition.

FT depicts a broad trend, and an extrapolation of historical rates therefore tends to underestimate future BAU deforestation for counties at the early stages in the transition (HFLD), while it tends to overestimate BAU deforestation for countries at the later stages (LFHD and LFLD).

Countries with high forest cover can be expected to be at early stages of the FT. GDP per capita captures the stage in a country's economic development, which is linked to the pattern of natural resource use, including forests. The choice of forest cover and GDP per capita also fits well with the two key scenarios in the FT:

(i) a forest scarcity path, where forest scarcity triggers forces (for example, higher prices of forest products) that lead to forest cover stabilization; and

(ii) an economic development path, where new and better off-farm employment opportunities associated with economic growth (= increasing GDP per capita) reduce profitability of frontier agriculture and slows deforestation.

Historical Causes

Prehistory

The Carboniferous Rainforest Collapse was an event that occurred 300 million years ago. Climate change devastated tropical rainforests causing the extinction of many plant and animal species. The change was abrupt, specifically, at this time climate became cooler and drier, conditions that are not favourable to the growth of rainforests and much of the biodiversity within them. Rainforests were fragmented forming shrinking 'islands' further and further apart. This sudden collapse affected several large groups, effects on amphibians were particularly devastating, while reptiles fared better, being ecologically adapted to the drier conditions that followed.

An array of Neolithic artifacts, including bracelets, axe heads, chisels, and polishing tools.

Rainforests once covered 14% of the earth's land surface; now they cover a mere 6% and experts estimate that the last remaining rainforests could be consumed in less than 40 years. Small scale deforestation was practiced by some societies for tens of thousands of years before the beginnings of civilization. The first evidence of deforestation appears in the Mesolithic period. It was probably used to convert closed forests into more open ecosystems favourable to game animals. With the advent of agriculture, larger areas began to be deforested, and fire became the prime tool to clear land for crops. In Europe there is little solid evidence before 7000 BC. Mesolithic foragers used fire to create openings for red deer and wild boar. In Great Britain, shade-tolerant species such as oak and ash are replaced in the pollen record by hazels, brambles, grasses and nettles. Removal of the forests led to decreased transpiration, resulting in the formation of upland peat bogs. Widespread decrease in elm pollen across Europe between 8400–8300 BC and 7200–7000 BC, starting in southern Europe and gradually moving north to Great Britain, may represent land clearing by fire at the onset of Neolithic agriculture.

The Neolithic period saw extensive deforestation for farming land. Stone axes were being made from about 3000 BC not just from flint, but from a wide variety of hard rocks from across Britain and North America as well. They include the noted Langdale axe industry in the English Lake District, quarries developed at Penmaenmawr in North Wales and numerous other locations. Rough-outs were made locally near the quarries, and some were polished locally to give a fine finish. This step not only increased the mechanical strength of the axe, but also made penetration of wood easier. Flint was still used from sources such as Grimes Graves but from many other mines across Europe.

Evidence of deforestation has been found in Minoan Crete; for example the environs of the Palace of Knossos were severely deforested in the Bronze Age.

Pre-industrial History

Throughout most of history, humans were hunter gatherers who hunted within forests. In most areas, such as the Amazon, the tropics, Central America, and the Caribbean, only after shortages of wood and other forest products occur are policies implemented to ensure forest resources are used in a sustainable manner.

In ancient Greece, Tjeered van Andel and co-writers summarized three regional studies of historic erosion and alluviation and found that, wherever adequate evidence exists, a major phase of erosion follows, by about 500-1,000 years the introduction of farming in the various regions of Greece, ranging from the later Neolithic to the Early Bronze Age. The thousand years following the mid-first millennium BC saw serious, intermittent pulses of soil erosion in numerous places. The historic silting of ports along the southern coasts of Asia Minor (*e.g.* Clarus, and the examples of Ephesus, Priene and Miletus, where harbors had to be abandoned because of the silt deposited by the Meander) and in coastal Syria during the last centuries BC.

Easter Island

Easter Island has suffered from heavy soil erosion in recent centuries, aggravated by agriculture and deforestation. Jared Diamond gives an extensive look into the collapse of the ancient Easter Islanders in his book *Collapse*. The disappearance of the island's trees seems to coincide with a decline of its civilization around the 17th and 18th century. He attributed the collapse to deforestation and over-exploitation of all resources.

The famous silting up of the harbor for Bruges, which moved port commerce to Antwerp, also followed a period of increased settlement growth (and apparently of deforestation) in the upper river basins. In early medieval Riez in upper Provence, alluvial silt from two small rivers raised the riverbeds and widened the floodplain, which slowly buried the Roman settlement in alluvium and gradually moved new construction to higher ground; concurrently the headwater valleys above Riez were being opened to pasturage.

A typical progress trap was that cities were often built in a forested area, which would provide wood for some industry (for example, construction, shipbuilding, pottery). When deforestation occurs without proper replanting, however; local wood supplies become difficult to obtain near enough to remain competitive, leading to the city's abandonment, as happened repeatedly in Ancient Asia Minor. Because of fuel needs, mining and metallurgy often led to deforestation and city abandonment.

With most of the population remaining active in (or indirectly dependent on) the agricultural sector, the main pressure in most areas remained land clearing for crop and cattle farming. Enough wild green was usually left standing (and partially used, for example, to collect firewood, timber and fruits, or to graze pigs) for wildlife to remain viable. The elite's (nobility and higher clergy) protection of their own hunting privileges and game often protected significant woodlands.

Major parts in the spread (and thus more durable growth) of the population were played by monastical 'pioneering' (especially by the Benedictine and Commercial orders) and some feudal lords' recruiting farmers to settle (and become tax payers) by offering relatively good legal and fiscal conditions. Even when speculators sought to encourage towns, settlers needed an agricultural belt around or sometimes within defensive walls. When populations were quickly decreased by causes such as the Black Death or devastating warfare (for example, Genghis Khan's Mongol hordes in eastern and central Europe, Thirty Years' War in Germany), this could lead to settlements being abandoned. The land was reclaimed by nature, but the secondary forests usually lacked the original biodiversity.

Deforestation of Brazil's Atlantic Forest c.1820-1825

From 1100 to 1500 AD, significant deforestation took place in Western Europe as a result of the expanding human population. The large-scale building of wooden sailing ships by European (coastal) naval owners since the 15th century for exploration, colonisation, slave trade–and other trade on the high seas consumed many forest resources. Piracy also contributed to the over harvesting of forests, as in Spain. This led to a weakening of the domestic economy after Columbus' discovery of America, as the economy became dependent on colonial activities (plundering, mining, cattle, plantations, trade, etc.)

In *Changes in the Land* (1983), William Cronon analyzed and documented 17th-century English colonists' reports of increased seasonal flooding in New England during the period when new settlers initially cleared the forests for agriculture. They believed flooding was linked to widespread forest clearing upstream.

The massive use of charcoal on an industrial scale in Early Modern Europe was a new type of consumption of western forests; even in Stuart England, the relatively primitive production of charcoal has already reached an impressive level. Stuart England was so widely deforested that it depended on the Baltic trade for ship timbers, and looked to the untapped forests of New England to supply the need. Each of Nelson's Royal Navy war ships at Trafalgar (1805) required 6,000 mature oaks for its construction. In France, Colbert planted oak forests to supply the French navy in the future. When the oak plantations matured in the mid-19th century, the masts were no longer required because shipping had changed.

Norman F. Cantor's summary of the effects of late medieval deforestation applies equally well to Early Modern Europe:

Europeans had lived in the midst of vast forests throughout the earlier medieval centuries. After 1250 they became so skilled at deforestation that by 1500 they were running short of wood for heating and cooking. They were faced with a nutritional decline because of the elimination of the generous supply of wild game that had inhabited the now-disappearing forests, which throughout medieval times had provided the staple of their carnivorous high-protein diet. By 1500 Europe was on the edge of a fuel and nutritional disaster [from] which it was saved in the sixteenth century only by the burning of soft coal and the cultivation of potatoes and maize.

Industrial Era

In the 19th century, introduction of steamboats in the United States was the cause of deforestation of banks of major rivers, such as the Mississippi River, with increased and more severe flooding one of the environmental results. The steamboat crews cut wood every day from the riverbanks to fuel the steam engines. Between St. Louis and the confluence with the Ohio River to the south, the Mississippi became more wide and shallow, and changed its channel laterally. Attempts to improve navigation by the use of snag pullers often resulted in crews' clearing large trees 100 to 200 feet (61 m) back from the banks. Several French colonial towns of the Illinois Country, such as Kaskaskia, Cahokia and St. Philippe, Illinois were flooded and abandoned in the late 19th century, with a loss to the cultural record of their archeology.

The wholescale clearance of woodland to create agricultural land can be seen in many parts of the world, such as the Central forest-grasslands transition and other areas of the Great Plains of the United States. Specific parallels are seen in the 20th-century deforestation occurring in many developing nations.

Rates of Deforestation

Global deforestation sharply accelerated around 1852. It has been estimated that about half of the Earth's mature tropical forests—between 7.5 million and 8 million km^2 (2.9 million to 3 million sq mi) of the original 15 million to 16 million km^2 (5.8 million to 6.2 million sq mi) that until 1947 covered the planet—have now been destroyed. Some scientists have predicted that unless significant measures (such as seeking out and protecting old growth forests that have not been disturbed) are taken on a worldwide basis, by 2030 there will only be 10% remaining, with another 10% in a degraded condition. 80% will have been lost, and with them hundreds of thousands of irreplaceable species. Some cartographers have attempted to illustrate the sheer scale of deforestation by country using a cartogram.

Slash-and-burn farming in the state of Rondônia, western Brazil

Estimates vary widely as to the extent of tropical deforestation. Scientists estimate that one fifth of the world's tropical rainforest was destroyed between 1960 and 1990. They claim that that rainforests 60 years ago covered 14% of the world's land surface, now only cover 5–7%, and that all tropical forests will be gone by the middle of the 21st century.

A 2002 analysis of satellite imagery suggested that the rate of deforestation in the humid tropics (approximately 5.8 million hectares per year) was roughly 23% lower than the most commonly quoted rates. Conversely, a newer analysis of satellite images reveals that deforestation of the Amazon rainforest is twice as fast as scientists previously estimated.

Some have argued that deforestation trends may follow a Kuznets curve, which if true would nonetheless fail to eliminate the risk of irreversible loss of non-economic forest values (for example, the extinction of species).

Satellite image of Haiti's border with the Dominican Republic shows the amount of deforestation on the Haitian side

A 2005 report by the United Nations Food and Agriculture Organization (FAO) estimates that although the Earth's total forest area continues to decrease at about 13 million hectares per year, the global rate of deforestation has recently been slowing. Still others claim that rainforests are being destroyed at an ever-quickening pace. The London-based Rainforest Foundation notes that "the UN figure is based on a definition of forest as being an area with as little as 10% actual tree cover, which would therefore include areas that are actually savannah-like ecosystems and badly damaged forests." Other critics of the FAO data point out that they do not distinguish between forest types, and that they are based largely on reporting from forestry departments of individual countries, which do not take into account unofficial activities like illegal logging.

Despite these uncertainties, there is agreement that destruction of rainforests remains a significant environmental problem. Up to 90% of West Africa's coastal rainforests have disappeared since 1900. In South Asia, about 88% of the rainforests have been lost. Much of what remains of the world's rainforests is in the Amazon basin, where the Amazon Rainforest covers approximately 4 million square kilometres. The regions with the highest tropical deforestation rate between 2000 and 2005 were Central America—which lost 1.3% of its forests each year—and tropical Asia. In Central America, two-thirds of lowland tropical forests have been turned into pasture since 1950 and 40% of all the rainforests have been lost in the last 40 years. Brazil has lost 90–95% of its Mata Atlântica forest. Paraguay was losing its natural semi humid forests in the country's western regions at a rate of 15.000 hectares at a randomly studied 2-month period in 2010, Paraguay's parliament refused in 2009 to pass a law that would have stopped cutting of natural forests altogether.

Deforestation around Pakke Tiger Reserve, India

Madagascar has lost 90% of its eastern rainforests. As of 2007, less than 1% of Haiti's forests remained. Mexico, India, the Philippines, Indonesia, Thailand, Burma, Malaysia, Bangladesh, China, Sri Lanka, Laos, Nigeria, the Democratic Republic of the Congo, Liberia, Guinea, Ghana and the Ivory Coast, have lost large areas of their rainforest. Several countries, notably Brazil, have declared their deforestation a national emergency. The World Wildlife Fund's ecoregion project catalogues habitat types throughout the world, including habitat loss such as deforestation, showing for example that even in the rich forests of parts of Canada such as the Mid-Continental Canadian forests of the prairie provinces half of the forest cover has been lost or altered.

Regions

Rates of deforestation vary around the world.

In 2011 Conservation International listed the top 10 most endangered forests, characterized by having all lost 90% or more of their original habitat, and each harboring at least 1500 endemic plant species (species found nowhere else in the world).

Top 10 Most Endangered Forests 2011				
Endangered forest	Region	Remaining habitat	Predominate vegetation type	Notes
Indo-Burma	Asia-Pacific	5%	Tropical and subtropical moist broadleaf forests	Rivers, floodplain wetlands, mangrove forests. Burma, Thailand, Laos, Vietnam, Cambodia, India.
New Caledonia	Asia-Pacific	5%	Tropical and subtropical moist broadleaf forests	
Sundaland	Asia-Pacific	7%	Tropical and subtropical moist broadleaf forests	Western half of the Indo-Malayan archipelago including southern Borneo and Sumatra.
Philippines	Asia-Pacific	7%	Tropical and subtropical moist broadleaf forests	Forests over the entire country including 7,100 islands.
Atlantic Forest	South America	8%	Tropical and subtropical moist broadleaf forests	Forests along Brazil's Atlantic coast, extends to parts of Paraguay, Argentina and Uruguay.
Mountains of Southwest China	Asia-Pacific	8%	Temperate coniferous forest	
California Floristic Province	North America	10%	Tropical and subtropical dry broadleaf forests	
Coastal Forests of Eastern Africa	Africa	10%	Tropical and subtropical moist broadleaf forests	Mozambique, Tanzania, Kenya, Somalia.
Madagascar & Indian Ocean Islands	Africa	10%	Tropical and subtropical moist broadleaf forests	Madagascar, Mauritius, Reunion, Seychelles, Comoros.
Eastern Afromontane	Africa	11%	Tropical and subtropical moist broadleaf forests Montane grasslands and shrublands	Forests scattered along the eastern edge of Africa, from Saudi Arabia in the north to Zimbabwe in the south.

Control

Reducing Emissions

Main international organizations including the United Nations and the World Bank, have begun to develop programs aimed at curbing deforestation. The blanket term Reducing Emissions from Deforestation and Forest Degradation (REDD) describes these sorts of programs, which use direct monetary or other incentives to encourage developing countries to limit and/or roll back deforestation. Funding has been an issue, but at the UN Framework Convention on Climate Change (UNFCCC) Conference of the Parties-15 (COP-15) in Copenhagen in December 2009, an accord was reached with a collective commitment by developed countries for new and additional resources, including forestry and investments through international institutions, that will approach USD 30 billion for the period 2010–2012. Significant work is underway on tools for use in monitoring developing country adherence to their agreed REDD targets. These tools, which rely on remote forest monitoring using satellite imagery and other data sources, include the Center for Global Development's FORMA (Forest Monitoring for Action) initiative and the Group on Earth Observations' Forest Carbon Tracking Portal. Methodological guidance for forest monitoring was also emphasized at COP-15. The environmental organization Avoided Deforestation Partners leads the campaign for development of REDD through funding from the U.S. government. In 2014, the Food and Agriculture Organization of the United Nations and partners launched Open Foris - a set of open-source software tools that assist countries in gathering, producing and disseminating information on the state of forest resources. The tools support the inventory lifecycle, from needs assessment, design, planning, field data collection and management, estimation analysis, and dissemination. Remote sensing image processing tools are included, as well as tools for international reporting for Reducing emissions from deforestation and forest degradation (REDD) and MRV and FAO's Global Forest Resource Assessments.

In evaluating implications of overall emissions reductions, countries of greatest concern are those categorized as High Forest Cover with High Rates of Deforestation (HFHD) and Low Forest Cover with High Rates of Deforestation (LFHD). Afghanistan, Benin, Botswana, Burma, Burundi, Cameroon, Chad, Ecuador, El Salvador, Ethiopia, Ghana, Guatemala, Guinea, Haiti, Honduras, Indonesia, Liberia, Malawi, Mali, Mauritania, Mongolia, Namibia, Nepal, Nicaragua, Niger, Nigeria, Pakistan, Paraguay, Philippines, Senegal, Sierra Leone, Sri Lanka, Sudan, Togo, Uganda, United Republic of Tanzania, Zimbabwe are listed as having Low Forest Cover with High Rates of Deforestation (LFHD). Brazil, Cambodia, Democratic Peoples Republic of Korea, Equatorial Guinea, Malaysia, Solomon Islands, Timor-Leste, Venezuela, Zambia are listed as High Forest Cover with High Rates of Deforestation (HFHD).

Payments for Conserving Forests

In Bolivia, deforestation in upper river basins has caused environmental problems, including soil erosion and declining water quality. An innovative project to try and remedy this situation involves landholders in upstream areas being paid by downstream water users to conserve forests. The landholders receive US$20 to conserve the trees, avoid polluting livestock practices, and enhance the biodiversity and forest carbon on their land. They also receive US$30, which purchases a beehive, to compensate for conservation for two hectares of water-sustaining forest for five years. Honey revenue per hectare of forest is US$5 per year, so within five years, the landholder has sold US$50 of honey. The project is being conducted by Fundación Natura Bolivia and Rare Conservation, with support from the Climate & Development Knowledge Network.

Farming

New methods are being developed to farm more intensively, such as high-yield hybrid crops, greenhouse, autonomous building gardens, and hydroponics. These methods are often dependent on chemical inputs to maintain necessary yields. In cyclic agriculture, cattle are grazed on farm land that is resting and rejuvenating. Cyclic agriculture actually increases the fertility of the soil. Intensive farming can also decrease soil nutrients by consuming at an accelerated rate the trace minerals needed for crop growth.The most promising approach, however, is the concept of food forests in permaculture, which consists of agroforestal systems carefully designed to mimic natural forests, with an emphasis on plant and animal species of interest for food, timber and other uses. These systems have low dependence on fossil fuels and agro-chemicals, are highly self-maintaining, highly productive, and with strong positive impact on soil and water quality, and biodiversity.

Monitoring Deforestation

There are multiple methods that are appropriate and reliable for reducing and monitoring deforestation. One method is the "visual interpretation of aerial photos or satellite imagery that is labor-intensive but does not require high-level training in computer image processing or extensive computational resources". Another method includes hot-spot analysis (that is, locations of rapid change) using expert opinion or coarse resolution satellite data to identify locations for detailed digital analysis with high resolution satellite images. Deforestation is typically assessed by quantifying the amount of area deforested, measured at the present time. From an environmental point of view, quantifying the damage and its possible consequences is a more important task, while conservation efforts are more focused on forested land protection and development of land-use alternatives to avoid continued deforestation. Deforestation rate and total area deforested, have been widely used for monitoring deforestation in many regions, including the Brazilian Amazon deforestation monitoring by INPE.

Forest Management

Efforts to stop or slow deforestation have been attempted for many centuries because it has long been known that deforestation can cause environmental damage sufficient in some cases to cause societies to collapse. In Tonga, paramount rulers developed policies designed to prevent conflicts between short-term gains from converting forest to farmland and long-term problems forest loss would cause, while during the 17th and 18th centuries in Tokugawa, Japan, the shoguns developed a highly sophisticated system of long-term planning to stop and even reverse deforestation of the preceding centuries through substituting timber by other products and more efficient use of land that had been farmed for many centuries. In 16th-century Germany, landowners also developed silviculture to deal with the problem of deforestation. However, these policies tend to be limited to environments with *good rainfall, no dry season* and *very young soils* (through volcanism or glaciation). This is because on older and less fertile soils trees grow too slowly for silviculture to be economic, whilst in areas with a strong dry season there is always a risk of forest fires destroying a tree crop before it matures.

In the areas where "slash-and-burn" is practiced, switching to "slash-and-char" would prevent the rapid deforestation and subsequent degradation of soils. The biochar thus created, given back to the soil, is not only a durable carbon sequestration method, but it also is an extremely beneficial amendment to the soil. Mixed with biomass it brings the creation of terra preta, one of the richest soils on the planet and the only one known to regenerate itself.

Sustainable Practices

Certification, as provided by global certification systems such as Programme for the Endorsement of Forest Certification and Forest Stewardship Council, contributes to tackling deforestation by creating market demand for timber from sustainably managed forests. According to the United Nations Food and Agriculture Organization (FAO), "A major condition for the adoption of sustainable forest management is a demand for products that are produced sustainably and consumer willingness to pay for the higher costs entailed. Certification represents a shift from regulatory approaches to market incentives to promote sustainable forest management. By promoting the positive attributes of forest products from sustainably managed forests, certification focuses on the demand side of environmental conservation." Rainforest Rescue argues that the standards of organizations like FSC are too closely connected to timber industry interests and therefore do not guarantee environmentally and socially responsible forest management. In reality, monitoring systems are inadequate and various cases of fraud have been documented worldwide.

Bamboo is advocated as a more sustainable alternative for cutting down wood for fuel.

Some nations have taken steps to help increase the amount of trees on Earth. In 1981, China created National Tree Planting Day Forest and forest coverage had now reached 16.55% of China's land mass, as against only 12% two decades ago

Using fuel from bamboo rather than wood results in cleaner burning, and since bamboo matures much faster than wood, deforestation is reduced as supply can be replenished faster.

Reforestation

In many parts of the world, especially in East Asian countries, reforestation and afforestation are increasing the area of forested lands. The amount of woodland has increased in 22 of the world's 50 most forested nations. Asia as a whole gained 1 million hectares of forest between 2000 and 2005. Tropical forest in El Salvador expanded more than 20% between 1992 and 2001. Based on these trends, one study projects that global forest will increase by 10%—an area the size of India— by 2050.

In the People's Republic of China, where large scale destruction of forests has occurred, the government has in the past required that every able-bodied citizen between the ages of 11 and 60 plant three to five trees per year or do the equivalent amount of work in other forest services. The government claims that at least 1 billion trees have been planted in China every year since 1982. This is no longer required today, but March 12 of every year in China is the Planting Holiday. Also, it has introduced the Green Wall of China project, which aims to halt the expansion of the Gobi desert through the planting of trees. However, due to the large percentage of trees dying off after planting (up to 75%), the project is not very successful. There has been a 47-million-hectare increase in forest area in China since the 1970s. The total number of trees amounted to be about 35 billion and 4.55% of China's land mass increased in forest coverage. The forest coverage was 12% two decades ago and now is 16.55%.

An ambitious proposal for China is the Aerially Delivered Re-forestation and Erosion Control System and the proposed Sahara Forest Project coupled with the Seawater Greenhouse.

In Western countries, increasing consumer demand for wood products that have been produced and harvested in a sustainable manner is causing forest landowners and forest industries to become increasingly accountable for their forest management and timber harvesting practices.

The Arbor Day Foundation's Rain Forest Rescue program is a charity that helps to prevent deforestation. The charity uses donated money to buy up and preserve rainforest land before the lumber companies can buy it. The Arbor Day Foundation then protects the land from deforestation. This also locks in the way of life of the primitive tribes living on the forest land. Organizations such as Community Forestry International, Cool Earth, The Nature Conservancy, World Wide Fund for Nature, Conservation International, African Conservation Foundation and Greenpeace also focus on preserving forest habitats. Greenpeace in particular has also mapped out the forests that are still intact and published this information on the internet. World Resources Institute in turn has made a simpler thematic map showing the amount of forests present just before the age of man (8000 years ago) and the current (reduced) levels of forest. These maps mark the amount of afforestation required to repair the damage caused by people.

Forest Plantations

To meet the world's demand for wood, it has been suggested by forestry writers Botkins and Sedjo that high-yielding forest plantations are suitable. It has been calculated that plantations yielding 10 cubic meters per hectare annually could supply all the timber required for international trade on 5% of the world's existing forestland. By contrast, natural forests produce about 1–2 cubic meters per hectare; therefore, 5–10 times more forestland would be required to meet demand. Forester Chad Oliver has suggested a forest mosaic with high-yield forest lands interspersed with conservation land.

In the country of Senegal, on the western coast of Africa, a movement headed by youths has helped to plant over 6 million mangrove trees. The trees will protect local villages from storm damages and will provide a habitat for local wildlife. The project started in 2008, and already the Senegalese government has been asked to establish rules and regulations that would protect the new mangrove forests.

Military Context

American Sherman tanks knocked out by Japanese artillery on Okinawa.

While the preponderance of deforestation is due to demands for agricultural and urban use for the human population, there are some examples of military causes. One example of deliberate deforestation is that which took place in the U.S. zone of occupation in Germany after World War II. Before the onset of the Cold War, defeated Germany was still considered a potential future threat rather than potential future ally. To address this threat, attempts were made to lower German industrial potential, of which forests were deemed an element. Sources in the U.S. government admitted that the purpose of this was that the "ultimate destruction of the war potential of German forests." As a consequence of the practice of clear-felling, deforestation resulted which could "be replaced only by long forestry development over perhaps a century."

Deforestation can also be one consequence of war. For example, in the 1945 Battle of Okinawa, bombardment and other combat operations reduced the lush tropical landscape into "a vast field of mud, lead, decay and maggots". Deforestation can also be an intentional tactic of military forces. Defoliants (like Agent Orange or others) was used by the British in the Malayan Emergency, and by the United States in the Korean War and Vietnam War.

Public Health Context

Deforestation eliminates a great number of species of plants and animals which also often results in an increase in disease. Loss of native species allows new species to come to dominance. Often the destruction of predatory species can result in an increase in rodent populations. These are known to carry plagues. Additionally, erosion can produce pools of stagnant water that are perfect breeding grounds for mosquitos, well known vectors of malaria, yellow fever, nipah virus, and more. Deforestation can also create a path for non-native species to flourish such as certain types of snails, which have been correlated with an increase in schistosomiasis cases.

Deforestation is occurring all over the world and has been coupled with an increase in the occurrence of disease outbreaks. In Malaysia, thousands of acres of forest have been cleared for pig farms. This has resulted in an increase in the zoonosis the Nipah virus. In Kenya, deforestation has led to an increase in malaria cases which is now the leading cause of morbidity and mortality the country.

Another pathway through which deforestation affects disease is the relocation and dispersion of disease-carrying hosts. This disease emergence pathway can be called "range expansion," whereby the host's range (and thereby the range of pathogens) expands to new geographic areas. Through deforestation, hosts and reservoir species are forced into neighboring habitats. Accompanying the reservoir species are pathogens that have the ability to find new hosts in previously unexposed regions. As these pathogens and species come into closer contact with humans, they are infected both directly and indirectly.

A catastrophic example of range expansion is the 1998 outbreak of Nipah Virus in Malaysia. For a number of years, deforestation, drought, and subsequent fires led to a dramatic geographic shift and density of fruit bats, a reservoir for Nipah virus. Deforestation reduced the available fruiting trees in the bats' habitat, and they encroached on surrounding orchards which also happened to be the location of a large number of pigsties. The bats, through proximity spread the Nipah to pigs. While the virus infected the pigs, mortality was much lower than among humans, making the pigs a virulent host leading to the transmission of the virus to humans. This resulted in 265 reported cases of encephalitis, of which 105 resulted in death. This example provides an important lesson for the impact deforestation can have on human health.

Another example of range expansion due to deforestation and other anthropogenic habitat impacts includes the Capybara rodent in Paraguay. This rodent is the host of a number of zoonotic diseases and, while there has not yet been a human-borne outbreak due to the movement of this rodent into new regions, it offers an example of how habitat destruction through deforestation and subsequent movements of species is occurring regularly.

A now well-developed theory is that the spread of HIV it is at least partially due deforestation. Rising populations created a food demand and with deforestation opening up new areas of the forest the hunters harvested a great deal of primate bushmeat, which is believed to be the origin of HIV.

Indirect Land use Change Impacts of Biofuels

The indirect land use change impacts of biofuels, also known as ILUC, relates to the unintended consequence of releasing more carbon emissions due to land-use changes around the world induced by the expansion of croplands for ethanol or biodiesel production in response to the increased global demand for biofuels.

Brazilian cerrado.

As farmers worldwide respond to higher crop prices in order to maintain the global food supply-and-demand balance, pristine lands are cleared to replace the food crops that were diverted elsewhere to biofuels' production. Because natural lands, such as rainforests and grasslands, store carbon in their soil and biomass as plants grow each year, clearance of wilderness for new farms translates to a net increase in greenhouse gas emissions. Due to this change in the carbon stock of the soil and the biomass, indirect land use change has consequences in the GHG balance of a biofuel.

Other authors have also argued that indirect land use changes produce other significant social and environmental impacts, affecting biodiversity, water quality, food prices and supply, land tenure, worker migration, and community and cultural stability.

Amazon rainforest.

History

The estimates of carbon intensity for a given biofuel depend on the assumptions regarding several variables. As of 2008, multiple full life cycle studies had found that corn ethanol, cellulosic ethanol and Brazilian sugarcane ethanol produce lower greenhouse gas emissions than gasoline. None of these studies, however, considered the effects of indirect land-use changes, and though land use impacts were acknowledged, estimation was considered too complex and difficult to model. A controversial paper published in February 2008 in Sciencexpress by a team led by Searchinger from Princeton University concluded that such effects offset the (positive) direct effects of both corn and cellulosic ethanol and that Brazilian sugarcane performed better, but still resulted in a small carbon debt.

Malaysian cloud forest

After the Searchinger team paper, estimation of carbon emissions from ILUC, together with the food vs. fuel debate, became one of the most contentious issues relating to biofuels, debated in the popular media, scientific journals, op-eds and public letters from the scientific community, and the ethanol industry, both American and Brazilian. This controversy intensified in April 2009 when the California Air Resources Board (CARB) set rules that included ILUC impacts to establish the California Low-Carbon Fuel Standard that entered into force in 2011.

In May 2009 U.S. Environmental Protection Agency (EPA) released a notice of proposed rulemaking for implementation of the 2007 modification of the Renewable Fuel Standard (RFS). EPA's proposed regulations also included ILUC, causing additional controversy among ethanol producers. EPA's February 3, 2010 final rule incorporated ILUC based on modelling that was significantly improved over the initial estimates.

The UK Renewable Transport Fuel Obligation program requires the Renewable Fuels Agency (RFA) to report potential indirect impacts of biofuel production, including indirect land use change or changes to food and other commodity prices. A July 2008 RFA study, known as the Gallager Review, found several risks and uncertainties, and that the "quantification of GHG emissions from indirect land-use change requires subjective assumptions and contains considerable uncertainty", and required further examination to properly incorporate indirect effects into calculation methodologies. A similarly cautious approach was followed by the European Union. In December 2008 the European Parliament adopted more stringent sustainability criteria for biofuels and directed the European Commission to develop a methodology to factor in GHG emissions from indirect land use change.

Studies and Controversy

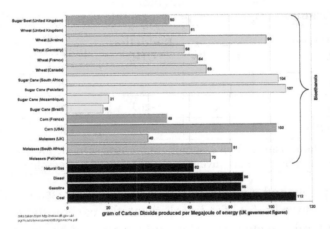

UK figures for the carbon intensity of bioethanol and fossil fuels. Graph assumes all bioethanols are burnt in country of origin and prior cropland was used to grow feedstock. No ILUC effects were included.

Before 2008, several full life cycle ("Well to Wheels" or WTW) studies had found that corn ethanol reduced transport-related greenhouse gas emissions. In 2007 a University of California, Berkeley team led by Farrel evaluated six previous studies, concluding that corn ethanol reduced GHG emissions by only 13 percent. However, 20 to 30 percent reduction for corn ethanol, and 85 to 85 percent for cellulosic ethanol, both figures estimated by Wang from Argonne National Laboratory, are more commonly cited. Wang reviewed 22 studies conducted between

1979 and 2005, and ran simulations with Argonne's GREET model. These studies accounted for direct land use changes. Several studies of Brazilian sugarcane ethanol showed that sugarcane as feedstock reduces GHG by 86 to 90 percent given no significant land use change. Estimates of carbon intensity depend on crop productivity, agricultural practices, power sources for ethanol distilleries and the energy efficiency of the distillery. None of these studies considered ILUC, due to estimation difficulties. Preliminary estimates by Delucchi from the University of California, Davis, suggested that carbon released by new lands converted to agricultural use was a large percentage of life-cycle emissions.

Searchinger and Fargione Studies

In 2008 Timothy Searchinger, a lawyer from Environmental Defense Fund, concluded that ILUC affects the life cycle assessment and that instead of saving, both corn and cellulosic ethanol increased carbon emissions as compared to gasoline by 93 and 50 percent respectively. Ethanol from Brazilian sugarcane performed better, recovering initial carbon emissions in 4 years, while U.S. corn ethanol required 167 years and cellulosic ethanol required a 52 years payback period. The study limited the analysis a 30-year period, assuming that land conversion emits 25 percent of the carbon stored in soils and all carbon in plants cleared for cultivation. Brazil, China, and India were considered among the overseas locations where land use change would occur as a result of diverting U.S. corn cropland, and it was assumed that new cropland in each of these regions correspond to different types of forest, savanna or grassland based on the historical proportion of each converted to cultivation in these countries during the 1990s.

Summary of Searchinger et al. comparison of corn ethanol and gasoline GHG emissions with and without land use change (Grams of CO_2 released per megajoule of energy in fuel)				
Fuel type (U.S.)	Carbon intensity	Reduction GHG	Carbon intensity + ILUC	Reduction GHG
Gasoline	92	-	92	-
Corn ethanol	74	-20%	177	+93%
Cellulosic ethanol	28	-70%	138	+50%
Notes: Calculated using default assumptions for 2015 scenario for ethanol in E85. Gasoline is a combination of conventional and reformulated gasoline.				

Fargione and his team published a separate paper in the same issue of Sciencexpress claiming that clearing lands to produce biofuel feedstock created a carbon deficit. This deficit applies to both direct and indirect land use changes. The study examined six conversion scenarios: Brazilian Amazon to soybean biodiesel, Brazilian Cerrado to soybean biodiesel, Brazilian Cerrado to sugarcane ethanol, Indonesian or Malaysian lowland tropical rainforest to palm biodiesel, Indonesian or Malaysian peatland tropical rainforest to palm biodiesel, and U.S. Central grassland to corn ethanol. The carbon debt was defined as the amount of CO_2 released during the first 50 years of this process of land conversion. For the two most common ethanol feedstocks, the study found that sugarcane ethanol produced on natural cerrado lands would take about 17 years to repay its carbon debt, while corn ethanol produced on U.S. central grasslands would result in a repayment time of about 93 years. The worst-case scenario is converting Indonesian or Malaysian tropical peatland rainforest to palm biodiesel production, which would require about 420 years to repay.

Criticism and Controversy

The Searchinger and Fargione studies created controversy in both the popular media and in scientific journals. Robert Zubrin observed that Searchinger's "indirect analysis" approach is pseudo-scientific and can be used to "prove anything".

Wang and Haq from Argonne National Laboratory claiming: the assumptions were outdated; they ignored the potential of increased efficiency; and no evidence showed that "U.S. corn ethanol production has so far caused indirect land use in other countries." They concluded that Searchinger demonstrated that ILUC "is much more difficult to model than direct land use changes". In his response Searchinger rebutted each technical objection and asserted that "... any calculation that ignores these emissions, however challenging it is to predict them with certainty, is too incomplete to provide a basis for policy decisions."

Another criticism, by Kline and Dale from Oak Ridge National Laboratory, held that Searchinger et al. and Fargione et al. "... do not provide adequate support for their claim that biofuels cause high emissions due to land-use change", as their conclusions depends on a misleading assumption because more comprehensive field research found that these land use changes "... are driven by interactions among cultural, technological, biophysical, economic, and demographic forces within a spatial and temporal context rather than by a single crop market". Fargione et al. responded in part that although many factors contributed to land clearing, this "observation does not diminish the fact that biofuels also contribute to land clearing if they are produced on existing cropland or on newly cleared lands". Searching disagreed with all of Kline and Dale arguments.

The U.S. biofuel industry also reacted, claiming that the "Searchinger study is clearly a 'worst case scenario' analysis ..." and that this study "relies on a long series of highly subjective assumptions ..." Searchinger rebutted each claim, concluding that NFA's criticisms were invalid. He noted that even if some of his assumptions are high estimates, the study also made many conservative assumptions.

Slash and burn forest removal in Brazil.

Cattle ranching in Brazil.

Brazil

In February 2010, Lapola estimated that planned expansion of Brazilian sugarcane and soybean biofuel plantations through 2020 would replace rangeland, with small direct land-use impact on carbon emissions. However, the expansion of the rangeland frontier into Amazonian forests, driven by cattle ranching, would indirectly offset the savings. "Sugarcane ethanol and soybean biodiesel each contribute to nearly half of the projected indirect deforestation of 121,970 km² by 2020, creating a carbon debt that would take about 250 years to be repaid...."

The research also found that oil palm would cause the least land-use changes and associated carbon debt. The analysis also modeled livestock density increases and found that "a higher increase of 0.13 head per hectare in the average livestock density throughout the country could avoid the indirect land-use changes caused by biofuels (even with soybean as the biodiesel feedstock), while still fulfilling all food and bioenergy demands." The authors conclude that intensification of cattle ranching and concentration on oil palm are required to achieve effective carbon savings, recommending closer collaboration between the biofuel and cattle-ranching sectors.

The main Brazilian ethanol industry organization (UNICA) commented that such studies missed the continuing intensification of cattle production already underway.

A study by Arima *et al.* published in May 2011 used spatial regression modeling to provide the first statistical assessment of ILUC for the Brazilian Amazon due to soy production. Previously, the indirect impacts of soy crops were only anecdotal or analyzed through demand models at a global scale, while the study took a regional approach. The analysis showed a strong signal linking the expansion of soybean fields in settled agricultural areas at the southern and eastern rims of the Amazon basin to pasture encroachments for cattle production on the forest frontier. The results demonstrate the need to include ILUC in measuring the carbon footprint of soy crops, whether produced for biofuels or other end-uses.

The Arima study is based on 761 municipalities located in the Legal Amazon of Brazil, and found that between 2003 and 2008, soybean areas expanded by 39,100 km² in the basin's agricultural areas, mainly in Mato Grosso. The model showed that a 10% (3,910 km²) reduction of soy in old pasture areas would have led to a reduction in deforestation of up to 40% (26,039 km²) in heavily forested municipalities of the Brazilian Amazon. The analysis showed that the displacement of cattle production due to agricultural expansion drives land use change in municipalities located hundreds of kilometers away, and that the Amazonian ILUC is not only measurable but its impact is significant.

Implementation

California LCFS

California carbon intensity values for gasoline, diesel and fuels that substitute them (grams of CO_2 equivalent released per MJ of energy produced)			
Fuel type	Carbon intensity	Carbon intensity + land-use changes	Intensity change respect to 2011 LCFS
Midwest corn ethanol	75.10	105.10	+10%
California gasoline	95.86	95.86	+0.2%
CARB LCFS 2011 for gasoline	-	95.61	-
California diesel (ULSD)	94.71	94.71	+0.2%
CARB LCFS 2011 for diesel	-	94.47	-
California ethanol	50.70	80.70	-16%
Brazilian sugarcane ethanol	27.40	73.40	-23%
Biodiesel (B100) Midwest soybeans	26.93	68.93	-27%
Renewable diesel Midwest soybeans	28.80	68.93	-27%
Cellulosic ethanol (farmed trees)	2.40	20.40	-79%
Compressed natural gas (bio-methane)	11.26	11.26	-88%
Note: The complete lifecycle analysis for these fuels and others evaluated are available at CARB's website. Preliminary values of fuels not included in the 2009 LCFS ruling and subject to refining.			

On April 23, 2009, California Air Resources Board (CARB) approved the specific rules and carbon intensity reference values for the California Low-Carbon Fuel Standard (LCFS) that take effect January 1, 2011. CARB's rulemaking included ILUC. For some biofuels, CARB identified land use changes as a significant source of additional GHG emissions. It established one standard for gasoline and alternative fuels, and a second for diesel fuel and its replacements.

Controversy

The public consultation process before the ruling, and the ruling itself were controversial, yielding 229 comments. ILUC was one of the most contentious issues. On June 24, 2008, 27 scientists and researchers submitted a letter saying, "As researchers and scientists in the field of biomass to biofuel conversion, we are convinced that there simply is not enough hard empirical data to base any sound policy regulation in regards to the indirect impacts of renewable biofuels production. The field is relative new, especially when compared to the vast knowledge base present in fossil fuel production, and the limited analyses are driven by assumptions that sometimes lack robust empirical validation." The New Fuels Alliance, representing more than two-dozen biofuel companies, researchers and investors, questioned the Board intention to include indirect land use change effects into account, wrote "While it is likely true that zero is not the right number for the indirect effects of any product in the real world, enforcing indirect effects in a piecemeal way could have very serious consequences for the LCFS.... The argument that zero is not the right number does not justify enforcing a different wrong number, or penalizing one fuel for one category of indirect effects while giving another fuel pathway a free pass."

On the other side, more than 170 scientists and economists urged that CARB, "include indirect land use change in the lifecycle analyses of heat-trapping emissions from biofuels and other transportation fuels. This policy will encourage development of sustainable, low-carbon fuels that avoid conflict with food and minimize harmful environmental impacts.... There are uncertainties inherent in estimating the magnitude of indirect land use emissions from biofuels, but assigning a value of zero is clearly not supported by the science."

Industry representatives complained that the final rule overstated the environmental effects of corn ethanol, and also criticized the inclusion of ILUC as an unfair penalty to domestic corn ethanol because deforestation in the developing world was being tied to U.S. ethanol production. The 2011 limit for LCFS means that Mid-west corn ethanol failed, unless current carbon intensity was reduced. Oil industry representatives complained that the standard left oil refiners with few options, such as Brazilian sugarcane ethanol, with its accompanying tariff. CARB officials and environmentalists counter that time and economic incentives will allow produces to adapt.

UNICA welcomed the ruling, while urging CARB to better reflect Brazilian practices, lowering their estimates of Brazilian emissions.

The only Board member who voted against the ruling explained that he had a "hard time accepting the fact that we're going to ignore the comments of 125 scientists", referring to the letter submitted by a group of scientists questioning the ILUC penalty. "They said the model was not good enough ... to use at this time as a component part of such an historic new standard." CARB advanced the expected date for an expert working group to report on ILUC with refined estimates from January 2012 to January 2011.

On December 2009 the Renewable Fuels Association (RFA) and Growth Energy, two U.S. ethanol lobbying groups, filed a lawsuit challenging LCFS' constitutionality. The two organizations argued that LCFS violated both the Supremacy Clause and the Commerce Clause, jeopardizing the nation-wide ethanol market.

EPA Renewable Fuel Standard

U.S. Environmental Protection Agency Draft life cycle GHG emissions reduction results for different time horizon and discount rate approaches (includes indirect land use change effects)		
Fuel Pathway	100 years + 2% discount rate	30 years + 0% discount rate
Corn ethanol (natural gas dry mill)[1]	-16%	+5%
Corn ethanol (Best case NG DM)[2]	-39%	-18%

Corn ethanol (coal dry mill)	+13%	+34%
Corn ethanol (biomass dry mill)	-39%	-18%
Corn ethanol (biomass dry mill with combined heat and power)	-47%	-26%
Soybean-based biodiesel	-22%	+4%
Waste grease biodiesel	-80%	-80%
Sugarcane ethanol	-44%	-26%
Cellulosic ethanol from switchgrass	-128%	-124%
Cellulosic ethanol from corn stover	-115%	-116%
Notes: (1) Dry mill (DM) plants grind the entire kernel and generally produce only one primary co-product: distillers grains with solubles (DGS). (2) Best case plants produce wet distillers grains co-product.		

The Energy Independence and Security Act of 2007 (EISA) established new renewable fuel categories and eligibility requirements, setting mandatory lifecycle emissions limits. EISA explicitly mandated EPA to include "direct emissions and significant indirect emissions such as significant emissions from land use changes."

EISA required a 20% reduction in lifecycle GHG emissions for any fuel produced at facilities that commenced construction after December 19, 2007 to be classified as a "renewable fuel"; a 50% reduction for fuels to be classified as "biomass-based diesel" or "advanced biofuel", and a 60% reduction to be classified as "cellulosic biofuel". EISA provided limited flexibility to adjust these thresholds downward by up to 10 percent, and EPA proposed this adjustment for the advanced biofuels category. Existing plants were grandfathered in.

On May 5, 2009, EPA released a notice of proposed rulemaking for implementation of the 2007 modification of the Renewable Fuel Standard, known as RFS2. The draft of the regulations was released for public comment during a 60-day period, a public hearing was held on 9 June 2009, and also a workshop was conducted on 10–11 June 2009.

EPA's draft analysis stated that ILUC can produce significant near-term GHG emissions due to land conversion, but that biofuels can pay these back over subsequent years. EPA highlighted two scenarios, varying the time horizon and the discount rate for valuing emissions. The first assumed a 30-year time period uses a 0 percent discount rate (valuing emissions equally regardless of timing). The second scenario used a 100-year time period and a 2% discount rate.

Maize is the main feedstock for the production of ethanol fuel in the U.S..

On the same day that EPA published its notice of proposed rulemaking, President Obama signed a Presidential Directive seeking to advance biofuels research and commercialization. The Directive established the Biofuels Interagency Working Group, to develop policy ideas for increasing investment in next-generation fuels and for reducing their environmental footprint.

Sugarcane is the main feedstock for the production of ethanol fuel in Brazil.

The inclusion of ILUC in the proposed ruling provoked complaints from ethanol and biodiesel producers. Several environmental organizations welcomed the inclusion of ILUC but criticized the consideration of a 100-year payback scenario, arguing that it underestimated land conversion effects. American corn growers, biodiesel producers, ethanol producers and Brazilian sugarcane ethanol producers complained about EPA's methodology, while the oil industry requested an implementation delay.

On June 26, 2009, the House of Representatives approved the American Clean Energy and Security Act 219 to 212, mandating EPA to exclude ILUC for a 5-year period, vis a vis RFS2. During this period, more research is to be conducted to develop more reliable models and methodologies for estimating ILUC, and Congress will review this issue before allowing EPA to rule on this matter. The bill failed in the U.S. Senate.

On February 3, 2010, EPA issued its final RFS2 rule for 2010 and beyond. The rule incorporated direct and significant indirect emissions including ILUC. EPA incorporated comments and data from new studies. Using a 30-year time horizon and a 0% discount rate, EPA concluded that multiple biofuels would meet this standard.

EPA's analysis accepted both ethanol produced from corn starch and biobutanol from corn starch as "renewable fuels". Ethanol produced from sugarcane became an "advanced fuel". Both diesel produced from algal oils and biodiesel from soy oil and diesel from waste oils, fats, and greases fell in the "biomass-based diesel" category. Cellulosic ethanol and cellulosic diesel met the "cellulosic biofuel" standard.

The table summarizes the mean GHG emissions estimated by EPA modelling and the range of variations considering that the main source of uncertainty in the life cycle analysis is the GHG emissions related to international land use change.

U.S. Environmental Protection Agency Life cycle Year 2022 GHG emissions reduction results for RFS2 final rule (includes direct and indirect land use change effects and a 30-year payback period at a 0% discount rate)			
Renewable fuel Pathway (for U.S. consumption)	Mean GHG emission reduction[1]	GHG emission reduction 95% confidence interval[2]	Assumptions/comments
Corn ethanol	21%	7–32%	New or expanded natural gas fired dry mill plant, 37% wet and 63% dry DGS it produces, and employing corn oil fractionation technology.
Corn biobutanol	31%	20–40%	Natural gas fired dry mill plant, 37% wet and 63% dry DGS it produces, and employing corn oil fractionation technology.
Sugarcane ethanol[3]	61%	52–71%	Ethanol is produced and dehydrated in Brazil prior to being imported into the U.S. and the residue is not collected. GHG emissions from ocean tankers hauling ethanol from Brazil to the U.S. are included.
Cellulosic ethanol from switchgrass	110%	102–117%	Ethanol produced using the biochemical process.
Cellulosic ethanol from corn stover	129%	No ILUC	Ethanol produced using the biochemical process. Ethanol produced from agricultural residues does not have any international land use emissions.
Biodiesel from soybean	57%	22–85%	Plant using natural gas.
Waste grease biodiesel	86%	No ILUC	Waste grease feedstock does not have any agricultural or land use emissions.

Notes: (1) Percent reduction in lifecycle GHG emissions compared to the average lifecycle GHG for gasoline or diesel sold or distributed as transportation fuel in 2005.

(2) Confidence range accounts for uncertainty in the types of land use change assumptions and the magnitude of resulting GHG emissions.

(3) A new Brazil module was develop to model the impact of increased production of Brazilian sugarcane ethanol for use in the U.S. market and the international impacts of Brazilian sugarcane ethanol production. The Brazil module also accounts for the domestic competition between crop and pasture land uses, and allows for livestock intensification (heads of cattle per unit area of land).

Reactions

UNICA welcomed the ruling, in particular, for the more precise lifecycle emissions estimate and hoped that classification the advanced biofuel designation would help eliminate the tariff.

The U.S. Renewable Fuels Association (RFA) also welcomed the ruling, as ethanol producers "require stable federal policy that provides them the market assurances they need to commercialize new technologies", restating their ILUC objection.

RFA also complained that corn-based ethanol scored only a 21% reduction, noting that without ILUC, corn ethanol achieves a 52% GHG reduction. RFA also objected that Brazilian sugarcane ethanol "benefited disproportionally" because EPA's revisions lowered the initially equal ILUC estimates by half for corn and 93% for sugarcane.

Several Midwestern lawmakers commented that they continued to oppose EPA's consideration of the "dicey science" of indirect land use that "punishes domestic fuels". House Agriculture Chairman Collin Peterson said, "... to think that we can credibly measure the impact of international indirect land use is completely unrealistic, and I will continue to push for legislation that prevents unreliable methods and unfair standards from burdening the biofuels industry."

EPA Administrator Lisa P. Jackson commented that the agency "did not back down from considering land use in its final rules, but the agency took new information into account that led to a more favorable calculation for ethanol". She cited new science and better data on crop yield and productivity, more information on co-products that could be produced from advanced biofuels and expanded land-use data for 160 countries, instead of the 40 considered in the proposed rule.

Europe

As of 2010, European Union and United Kingdom regulators had recognized the need to take ILUC into account, but had not determined the most appropriate methodology.

UK Renewable Transport Fuel Obligation

The UK Renewable Transport Fuel Obligation (RTFO) program requires fuel suppliers to report direct impacts, and asked the Renewable Fuels Agency (RFA) to report potential indirect impacts, including ILUC and commodity price changes. The RFA's July 2008 "Gallager Review", mentioned several risks regarding biofuels and required feedstock production to avoid agricultural land that would otherwise be used for food production, despite concluding that "quantification of GHG emissions from indirect land-use change requires subjective assumptions and contains considerable uncertainty". Some environmental groups argued that emissions from ILUC were not being taken into account and could be creating more emissions.

European Union

On December 17, 2008, the European Parliament approved the Renewable Energy Sources Directive (COM(2008)19) and amendments to the Fuel Quality Directive (Directive 2009/30), which included sustainability criteria for biofuels and mandated consideration of ILUC. The Directive established a 10% biofuel target. A separate Fuel Quality Directive set the EU's Low Carbon Fuel Standard, requiring a 6% reduction in GHG intensity of EU transport fuels by 2020. The legislation ordered the European Commission to develop a methodology to factor in GHG emissions from ILUC by December 31, 2010, based on the best available scientific evidence.

In the meantime, the European Parliament defined lands that were ineligible for producing biofuel feedstocks for the purpose of the Directives. This category included wetlands and continuously forested areas with canopy cover of more than 30 percent or cover between 10 and 30 percent given evidence that its existing carbon stock was low enough to justify conversion.

The Commission subsequently published terms of reference for three ILUC modeling exercises: one using a General Equilibrium model; one using a Partial Equilibrium model and one comparing other global modeling exercises. It also consulted on a limited range of high-level options for addressing ILUC to which 17 countries and 59 organizations responded. The United Nations

Special Rapporteur on the Right to Food and several environmental organizations complained that the 2008 safeguards were inadequate. UNICA called for regulators to establish an empirical and "globally accepted methodology" to consider ILUC, with the participation of researchers and scientists from biofuel crop-producing countries.

In 2010 some NGOs accused the European Commission of lacking transparency given its reluctance to release documents relating to the ILUC work. In March 2010 the Partial and General Equilibrium Modelling results were made available, with the disclaimer that the EC had not adopted the views contained in the materials. These indicate that a 1.25% increase in EU biofuel consumption would require around 5,000,000 hectares (12,000,000 acres) of land globally.

The scenarios for varied from 5.6-8.6% of road transport fuels. The study found that ILUC effects offset part of the emission benefits, and that above the 5.6% threshold, ILUC emissions increase rapidly increase. For the expected scenario of 5.6% by 2020, the study estimated that biodiesel production increases would be mostly domestic, while bioethanol production would take place mainly in Brazil, regardless of EU duties. The analysis concluded that eliminating trade barriers would further reduce emissions, because the EU would import more from Brazil. Under this scenario, "*direct emission savings from biofuels are estimated at 18 Mt CO2, additional emissions from ILUC at 5.3 Mt CO2 (mostly in Brazil), resulting in a global net balance of nearly 13 Mt CO2 savings in a 20 years horizon.* The study also found that ILUC emissions were much greater for biodiesel from vegetable oil and estimated that in 2020 even at the 5.6% level were over half the greenhouse gas emissions from diesel.

As part of the announcement, the Commission stated that it would publish a report on ILUC by the end of 2010.

Certification System

On June 10, 2010, the EC announced its decision to set up certification schemes for biofuels, including imports as part of the Renewable Energy Directive. The Commission encouraged E.U. nations, industry and NGOs to set up voluntary certification schemes. EC figures for 2007 showed that 26% of biodiesel and 31% of bioethanol used in the E.U. was imported, mainly from Brazil and the United States.

Reactions

UNICA welcomed the EU efforts to "engage independent experts in its assessments" but requested that improvements because "... the report currently contains a certain number of inaccuracies, so once these are corrected, we anticipate even higher benefits resulting from the use of Brazilian sugarcane ethanol." UNICA highlighted the fact that the report assumed land expansion that "does not take into consideration the agro-ecological zoning for sugarcane in Brazil, which prevents cane from expanding into any type of native vegetation."

Critics said the 10% figure was reduced to 5.6% of transport fuels partly by exaggerating the contribution of electric vehicles (EV) in 2020, as the study assumed EVs would represent 20% of new car sales, two and six times the car industry's own estimate. They also claimed the study "exaggerates to around 45 percent the contribution of bioethanol—the greenest of all biofuels—and consequently downplays the worst impacts of biodiesel."

Environmental groups found that the measures "are too weak to halt a dramatic increase in deforestation". According to Greenpeace, "indirect land-use change impacts of biofuel production still are not properly addressed", which for them was the most dangerous problem of biofuels

Industry representatives welcomed the certification system and some dismissed concerns regarding the lack of land use criteria. UNICA and other industry groups wanted the gaps in the rules filled to provide a clear operating framework.

The negotiations between the European Parliament and the Council of European Ministers continue. A deal is not foreseen before 2014

References

- Sperling, Daniel; Deborah Gordon (2009). "Two billion cars: driving toward sustainability". Oxford University Press, New York: 98–99. ISBN 978-0-19-537664-7.

- Goettemoeller, Jeffrey; Adrian Goettemoeller (2007). "Sustainable Ethanol: Biofuels, Biorefineries, Cellulosic Biomass, Flex-Fuel Vehicles, and Sustainable Farming for Energy Independence". Prairie Oak Publishing, Maryville, Missouri: 40–41. ISBN 978-0-9786293-0-4.

- Jim Lane (2009-02-24). "CARB votes 9-1 for California Low Carbon Fuel Standard; moves up indirect land use review to Jan 2011 in response to outcry on ILUC". BiofuelsDigest. Retrieved 2009-04-29.

- "Renewable Fuel Standard Program(RFS2) Regulatory Impact Analysis" (PDF). U.S. Environmental Protection Agency. February 2010. Archived from the original (PDF) on February 2, 2011. Retrieved 2010-02-12.

- "Indirect land use change – Possible elements of a policy approach – preparatory draft for stakeholder/expert comments" (PDF). European Commission. Retrieved 2010-01-29.

- International Food Policy Research Institute (IFPRI) (2010-03-25). "Global Trade and Environmental Impact Study of the EU Biofuels Mandate" (PDF). Directorate General for Trade of the European Commission. Retrieved 2010-03-29.

- "Analysis Finds That First-Generation Biofuel Use of Up to 5.6% in EU Road Transport Fuels Delivers Net GHG Emissions Benefits After Factoring in Indirect Land Use Change". Green Car Congress. 2010-03-26. Retrieved 2010-03-29.

- Inslee, Jay; Bracken Hendricks (2007). "6. Homegrown Energy". Apollo's Fire. Island Press, Washington, D.C. pp. 153–155, 160–161. ISBN 978-1-59726-175-3.

- Christine Moser; Tina Hildebrandt; Robert Bailis (14 November 2013). "International Sustainability Standards and Certification". In Barry D. Solomon. Sustainable Development of Biofuels in Latin America and the Caribbean. Robert Bailis. Springer New York. pp. 27–69. ISBN 978-1-4614-9274-0.

- Rudel, T.K. 2005 "Tropical Forests: Regional Paths of Destruction and Regeneration in the Late 20th Century" Columbia University Press ISBN 0-231-13195-X

- Timothy Charles Whitmore; Jeffrey Sayer; International Union for Conservation of Nature and Natural Resources. General Assembly; IUCN Forest Conservation Programme (15 February 1992). Tropical deforestation and species extinction. Springer. ISBN 978-0-412-45520-9. Retrieved 4 December 2011.

- Taylor, Leslie (2004). The Healing Power of Rainforest Herbs: A Guide to Understanding and Using Herbal Medicinals. Square One. ISBN 9780757001444.

- Ron Nielsen, The Little Green Handbook: Seven Trends Shaping the Future of Our Planet, Picador, New York (2006) ISBN 978-0-312-42581-4

- John F. Mongillo; Linda Zierdt-Warshaw (2000). Linda Zierdt-Warshaw, ed. Encyclopedia of environmental science. University of Rochester Press. p. 104. ISBN 978-1-57356-147-1.

- Daniel B. Botkin (2001). No man's garden: Thoreau and a new vision for civilization and nature. Island Press. pp. 246–247. ISBN 978-1-55963-465-6. Retrieved 4 December 2011.

- "Stolen Goods: The EU's complicity in illegal tropical deforestation" (PDF). Forests and the European Union Resource Network. March 17, 2015. Retrieved March 31, 2015.

- EPA, US EPA. "2014 Renewable Fuel Standards for Renewable Fuel Standard program (RFS2): Notice of Proposed Rulemaking". Renewable Fuels: Regulations & Standards. United States Environmental Protection Agency. Retrieved 15 November 2013.

- "Greenhouse Gas and Energy Life Cycle Assessment of Pine Chemicals Derived from Crude Tall Oil and Their Substitutes". conducted by Franklin Associates, a Division of Eastern Research Group. August 2013. Retrieved 2014-07-03.

- "A Policy Framework for Climate and Energy in the Period From 2020 to 2030". European Commission. January 22, 2014. Retrieved 2014-07-03.

- "Biofuels: Worried about competition, chemical makers jump into debate". Peterka, Amanda, Greenwire. February 13, 2014. Retrieved 2014-07-03.

- Wald, Matthew L. (17 November 2013). "For First Time, E.P.A. Proposes Reducing Ethanol Requirement for Gas Mix". New York Times. Retrieved 15 November 2013.

Study of Biofuel Production Mechanism

The production mechanism of biofuel is discussed in this chapter. Biofuel is derived from biomass and can be in solid, liquid or gaseous form. Waste management issues and pollution can be drastically reduced if we generate energy using biomass. The chapter strategically encompasses and incorporates the major components and key concepts of biofuel, providing a complete understanding.

Anaerobic Digestion

Anaerobic digestion is a collection of processes by which microorganisms break down biodegradable material in the absence of oxygen. The process is used for industrial or domestic purposes to manage waste and/or to produce fuels. Much of the fermentation used industrially to produce food and drink products, as well as home fermentation, uses anaerobic digestion.

Anaerobic digestion occurs naturally in some soils and in lake and oceanic basin sediments, where it is usually referred to as "anaerobic activity". This is the source of marsh gas methane as discovered by Volta in 1776.

The digestion process begins with bacterial hydrolysis of the input materials. Insoluble organic polymers, such as carbohydrates, are broken down to soluble derivatives that become available for other bacteria. Acidogenic bacteria then convert the sugars and amino acids into carbon dioxide, hydrogen, ammonia, and organic acids. These bacteria convert these resulting organic acids into acetic acid, along with additional ammonia, hydrogen, and carbon dioxide. Finally, methanogens convert these products to methane and carbon dioxide. The methanogenic archaea populations play an indispensable role in anaerobic wastewater treatments.

It is used as part of the process to treat biodegradable waste and sewage sludge. As part of an integrated waste management system, anaerobic digestion reduces the emission of landfill gas into the atmosphere. Anaerobic digesters can also be fed with purpose-grown energy crops, such as maize.

Anaerobic digestion is widely used as a source of renewable energy. The process produces a biogas, consisting of methane, carbon dioxide and traces of other 'contaminant' gases. This biogas can be used directly as fuel, in combined heat and power gas engines or upgraded to natural gas-quality biomethane. The nutrient-rich digestate also produced can be used as fertilizer.

With the re-use of waste as a resource and new technological approaches which have lowered capital costs, anaerobic digestion has in recent years received increased attention among governments in a number of countries, among these the United Kingdom (2011), Germany and Denmark (2011).

Process

Many microorganisms affect anaerobic digestion, including acetic acid-forming bacteria (ace-togens) and methane-forming archaea (methanogens). These organisms promote a number of chemical processes in converting the biomass to biogas.

Gaseous oxygen is excluded from the reactions by physical containment. Anaerobes utilize electron acceptors from sources other than oxygen gas. These acceptors can be the organic material itself or may be supplied by inorganic oxides from within the input material. When the oxygen source in an anaerobic system is derived from the organic material itself, the 'intermediate' end products are primarily alcohols, aldehydes, and organic acids, plus carbon dioxide. In the presence of special-ised methanogens, the intermediates are converted to the 'final' end products of methane, carbon dioxide, and trace levels of hydrogen sulfide. In an anaerobic system, the majority of the chemical energy contained within the starting material is released by methanogenic bacteria as methane.

Populations of anaerobic microorganisms typically take a significant period of time to establish themselves to be fully effective. Therefore, common practice is to introduce anaerobic microorgan-isms from materials with existing populations, a process known as "seeding" the digesters, typi-cally accomplished with the addition of sewage sludge or cattle slurry.

Process Stages

The four key stages of anaerobic digestion involve hydrolysis, acidogenesis, acetogenesis and methanogenesis. The overall process can be described by the chemical reaction, where organic material such as glucose is biochemically digested into carbon dioxide (CO_2) and methane (CH_4) by the anaerobic microorganisms.

$$C_6H_{12}O_6 \rightarrow 3CO_2 + 3CH_4$$

Hydrolysis

In most cases, biomass is made up of large organic polymers. For the bacteria in anaerobic di-gesters to access the energy potential of the material, these chains must first be broken down into their smaller constituent parts. These constituent parts, or monomers, such as sugars, are readily available to other bacteria. The process of breaking these chains and dissolving the smaller molecules into solution is called hydrolysis. Therefore, hydrolysis of these high-molec-ular-weight polymeric components is the necessary first step in anaerobic digestion. Through hydrolysis the complex organic molecules are broken down into simple sugars, amino acids, and fatty acids.

Acetate and hydrogen produced in the first stages can be used directly by methanogens. Other molecules, such as volatile fatty acids (VFAs) with a chain length greater than that of acetate must first be catabolised into compounds that can be directly used by methanogens.

Acidogenesis

The biological process of acidogenesis results in further breakdown of the remaining compo-nents by acidogenic (fermentative) bacteria. Here, VFAs are created, along with ammonia, carbon dioxide, and hydrogen sulfide, as well as other byproducts. The process of acidogenesis is similar to the way milk sours.

Acetogenesis

The third stage of anaerobic digestion is acetogenesis. Here, simple molecules created through the acidogenesis phase are further digested by acetogens to produce largely acetic acid, as well as carbon dioxide and hydrogen.

Methanogenesis

The terminal stage of anaerobic digestion is the biological process of methanogenesis. Here, methanogens use the intermediate products of the preceding stages and convert them into methane, carbon dioxide, and water. These components make up the majority of the biogas emitted from the system. Methanogenesis is sensitive to both high and low pHs and occurs between pH 6.5 and pH 8. The remaining, indigestible material the microbes cannot use and any dead bacterial remains constitute the digestate.

Configuration

Anaerobic digesters can be designed and engineered to operate using a number of different configurations and can be categorized into batch vs. continuous process mode, mesophilic vs. thermophilic temperature conditions, high vs. low portion of solids, and single stage vs. multistage processes. More initial build money and a larger volume of the batch digester is needed to handle the same amount of waste as a continuous process digester. Higher heat energy is demanded in a thermophilic system compared to a mesophilic system and has a larger gas output capacity and higher methane gas content. For solids content, low will handle up to 15% solid content. Above this level is considered high solids content and can also be known as dry digestion. In a single stage process, one reactor houses the four anaerobic digestion steps. A multistage process utilizes two or more reactors for digestion to separate the methanogenesis and hydrolysis phases.

Batch or Continuous

Anaerobic digestion can be performed as a batch process or a continuous process. In a batch system biomass is added to the reactor at the start of the process. The reactor is then sealed for the duration of the process. In its simplest form batch processing needs inoculation with already processed material to start the anaerobic digestion. In a typical scenario, biogas production will be formed with a normal distribution pattern over time. Operators can use this fact to determine when they believe the process of digestion of the organic matter has completed. There can be severe odour issues if a batch reactor is opened and emptied before the process is well completed. A more advanced type of batch approach has limited the odour issues by integrating anaerobic digestion with in-vessel composting. In this approach inoculation takes place through the use of recirculated degasified percolate. After anaerobic digestion has completed, the biomass is kept in the reactor which is then used for in-vessel composting before it is opened As the batch digestion is simple and requires less equipment and lower levels of design work, it is typically a cheaper form of digestion. Using more than one batch reactor at a plant can ensure constant production of biogas.

In continuous digestion processes, organic matter is constantly added (continuous complete mixed) or added in stages to the reactor (continuous plug flow; first in – first out). Here, the end products are constantly or periodically removed, resulting in constant production of biogas. A single or multiple digesters in sequence may be used. Examples of this form of anaerobic digestion include continuous stirred-tank reactors, upflow anaerobic sludge blankets, expanded granular sludge beds and internal circulation reactors.

Temperature

The two conventional operational temperature levels for anaerobic digesters determine the species of methanogens in the digesters:

- *Mesophilic* digestion takes place optimally around 30 to 38 °C, or at ambient temperatures between 20 and 45 °C, where mesophiles are the primary microorganism present.
- *Thermophilic* digestion takes place optimally around 49 to 57 °C, or at elevated temperatures up to 70 °C, where thermophiles are the primary microorganisms present.

A limit case has been reached in Bolivia, with anaerobic digestion in temperature working conditions of less than 10 °C. The anaerobic process is very slow, taking more than three times the normal mesophilic time process. In experimental work at University of Alaska Fairbanks, a 1,000 litre digester using psychrophiles harvested from "mud from a frozen lake in Alaska" has produced 200–300 litres of methane per day, about 20 to 30% of the output from digesters in warmer climates. Mesophilic species outnumber thermophiles, and they are also more tolerant to changes in environmental conditions than thermophiles. Mesophilic systems are, therefore, considered to be more stable than thermophilic digestion systems. In contrast, while thermophilic digestion systems are considered less stable, their energy input is higher, with more biogas being removed from the organic matter in an equal amount of time. The increased temperatures facilitate faster reaction rates, and thus faster gas yields. Operation at higher temperatures facilitates greater pathogen reduction of the digestate. In countries where legislation, such as the Animal By-Products Regulations in the European Union, requires digestate to meet certain levels of pathogen reduction there may be a benefit to using thermophilic temperatures instead of mesophilic.

Additional pre-treatment can be used to reduce the necessary retention time to produce biogas. For example, certain processes shred the substrates to increase the surface area or use a thermal pretreatment stage (such as pasteurisation) to significantly enhance the biogas output. The pasteurisation process can also be used to reduce the pathogenic concentration in the digesate leaving the anaerobic digester. Pasteurisation may be achieved by heat treatment combined with maceration of the solids.

Solids Content

In a typical scenario, three different operational parameters are associated with the solids content of the feedstock to the digesters:

- High solids (dry—stackable substrate)
- High solids (wet—pumpable substrate)
- Low solids (wet—pumpable substrate)

High solids (dry) digesters are designed to process materials with a solids content between 25 and 40%. Unlike wet digesters that process pumpable slurries, high solids (dry – stackable substrate) digesters are designed to process solid substrates without the addition of water. The primary styles of dry digesters are continuous vertical plug flow and batch tunnel horizontal digesters. Continuous vertical plug flow digesters are upright, cylindrical tanks where feedstock is continuously fed into the top of the digester, and flows downward by gravity during digestion. In batch tunnel digesters, the feedstock is deposited in tunnel-like chambers with a gas-tight door. Neither approach has mixing inside the digester. The amount of pretreatment, such as contaminant removal, depends both upon the nature of the waste streams being processed and the desired quality of the digestate. Size reduction (grinding) is beneficial in continuous vertical systems, as it accelerates digestion, while batch systems avoid grinding and instead require structure (e.g. yard waste) to reduce compaction of the stacked pile. Continuous vertical dry digesters have a smaller footprint due to the shorter effective retention time and vertical design. Wet digesters can be designed to operate in either a high-solids content, with a total suspended solids (TSS) concentration greater than ~20%, or a low-solids concentration less than ~15%.

High solids (wet) digesters process a thick slurry that requires more energy input to move and process the feedstock. The thickness of the material may also lead to associated problems with abrasion. High solids digesters will typically have a lower land requirement due to the lower volumes associated with the moisture. High solids digesters also require correction of conventional performance calculations (e.g. gas production, retention time, kinetics, etc.) originally based on very dilute sewage digestion concepts, since larger fractions of the feedstock mass are potentially convertible to biogas.

Low solids (wet) digesters can transport material through the system using standard pumps that require significantly lower energy input. Low solids digesters require a larger amount of land than high solids due to the increased volumes associated with the increased liquid-to-feedstock ratio of the digesters. There are benefits associated with operation in a liquid environment, as it enables more thorough circulation of materials and contact between the bacteria and their food. This enables the bacteria to more readily access the substances on which they are feeding, and increases the rate of gas production.

Complexity

Digestion systems can be configured with different levels of complexity. In a single-stage digestion system (one-stage), all of the biological reactions occur within a single, sealed reactor or holding tank. Using a single stage reduces construction costs, but results in less control of the reactions occurring within the system. Acidogenic bacteria, through the production of acids, reduce the pH of the tank. Methanogenic bacteria, as outlined earlier, operate in a strictly defined pH range. Therefore, the biological reactions of the different species in a single-stage reactor can be in direct competition with each other. Another one-stage reaction system is an anaerobic lagoon. These lagoons are pond-like, earthen basins used for the treatment and long-term storage of manures. Here the anaerobic reactions are contained within the natural anaerobic sludge contained in the pool.

In a two-stage digestion system (multistage), different digestion vessels are optimised to bring maximum control over the bacterial communities living within the digesters. Acidogenic bacteria produce organic acids and more quickly grow and reproduce than methanogenic bacteria. Methanogenic bacteria require stable pH and temperature to optimise their performance.

Under typical circumstances, hydrolysis, acetogenesis, and acidogenesis occur within the first reaction vessel. The organic material is then heated to the required operational temperature (either mesophilic or thermophilic) prior to being pumped into a methanogenic reactor. The initial hydrolysis or acidogenesis tanks prior to the methanogenic reactor can provide a buffer to the rate at which feedstock is added. Some European countries require a degree of elevated heat treatment to kill harmful bacteria in the input waste. In this instance, there may be a pasteurisation or sterilisation stage prior to digestion or between the two digestion tanks. Notably, it is not possible to completely isolate the different reaction phases, and often some biogas is produced in the hydrolysis or acidogenesis tanks.

Residence Time

The residence time in a digester varies with the amount and type of feed material, and with the configuration of the digestion system. In a typical two-stage mesophilic digestion, residence time varies between 15 and 40 days, while for a single-stage thermophilic digestion, residence times is normally faster and takes around 14 days. The plug-flow nature of some of these systems will mean the full degradation of the material may not have been realised in this timescale. In this event, digestate exiting the system will be darker in colour and will typically have more odour.

In the case of an upflow anaerobic sludge blanket digestion (UASB), hydraulic residence times can be as short as 1 hour to 1 day, and solid retention times can be up to 90 days. In this manner, a UASB system is able to separate solids and hydraulic retention times with the use of a sludge blanket. Continuous digesters have mechanical or hydraulic devices, depending on the level of solids in the material, to mix the contents, enabling the bacteria and the food to be in contact. They also allow excess material to be continuously extracted to maintain a reasonably constant volume within the digestion tanks.

Inhibition

The anaerobic digestion process can be inhibited by several compounds, affecting one or more of the bacterial groups responsible for the different organic matter degradation steps. The degree of the inhibition depends, among other factors, on the concentration of the inhibitor in the digester. Potential inhibitors are ammonia, sulfide, light metal ions (Na, K, Mg, Ca, Al), heavy metals, some organics (chlorophenols, halogenated aliphatics, N-substituted aromatics, long chain fatty acids), etc.

Farm-based maize silage digester located near Neumünster in Germany, 2007 - the green, inflatable biogas holder is shown on top of the digester. *Right:* Two-stage, low solids, UASB digestion component of a mechanical biological treatment system near Tel Aviv; the process water is seen in balance tank and sequencing batch reactor, 2005.

Feedstocks

Anaerobic lagoon and generators at the Cal Poly Dairy, United States

The most important initial issue when considering the application of anaerobic digestion systems is the feedstock to the process. Almost any organic material can be processed with anaerobic digestion; however, if biogas production is the aim, the level of putrescibility is the key factor in its successful application. The more putrescible (digestible) the material, the higher the gas yields possible from the system.

Feedstocks can include biodegradable waste materials, such as waste paper, grass clippings, leftover food, sewage, and animal waste. Woody wastes are the exception, because they are largely unaffected by digestion, as most anaerobes are unable to degrade lignin. Xylophalgeous anaerobes (lignin consumers) or using high temperature pretreatment, such as pyrolysis, can be used to break down the lignin. Anaerobic digesters can also be fed with specially grown energy crops, such as silage, for dedicated biogas production. In Germany and continental Europe, these facilities are referred to as "biogas" plants. A codigestion or cofermentation plant is typically an agricultural anaerobic digester that accepts two or more input materials for simultaneous digestion.

The length of time required for anaerobic digestion depends on the chemical complexity of the material. Material rich in easily digestible sugars breaks down quickly where as intact lignocellulosic material rich in cellulose and hemicellulose polymers can take much longer to break down. Anaerobic microorganisms are generally unable to break down lignin, the recalcitrant aromatic component of biomass.

Anaerobic digesters were originally designed for operation using sewage sludge and manures. Sewage and manure are not, however, the material with the most potential for anaerobic digestion, as the biodegradable material has already had much of the energy content taken out by the animals that produced it. Therefore, many digesters operate with codigestion of two or more types of feedstock. For example, in a farm-based digester that uses dairy manure as the primary feedstock, the gas production may be significantly increased by adding a second feedstock, e.g., grass and corn (typical on-farm feedstock), or various organic byproducts, such as slaughterhouse waste, fats, oils and grease from restaurants, organic household waste, etc. (typical off-site feedstock).

Digesters processing dedicated energy crops can achieve high levels of degradation and biogas production. Slurry-only systems are generally cheaper, but generate far less energy than those using crops, such as maize and grass silage; by using a modest amount of crop material (30%), an anaerobic digestion plant can increase energy output tenfold for only three times the capital cost, relative to a slurry-only system.

Moisture Content

A second consideration related to the feedstock is moisture content. Dryer, stackable substrates, such as food and yard waste, are suitable for digestion in tunnel-like chambers. Tunnel-style systems typically have near-zero wastewater discharge, as well, so this style of system has advantages where the discharge of digester liquids are a liability. The wetter the material, the more suitable it will be to handling with standard pumps instead of energy-intensive concrete pumps and physical means of movement. Also, the wetter the material, the more volume and area it takes up relative to the levels of gas produced. The moisture content of the target feedstock will also affect what type of system is applied to its treatment. To use a high-solids anaerobic digester for dilute feedstocks, bulking agents, such as compost, should be applied to increase the solids content of the input material. Another key consideration is the carbon:nitrogen ratio of the input material. This ratio is the balance of food a microbe requires to grow; the optimal C:N ratio is 20–30:1. Excess N can lead to ammonia inhibition of digestion.

Contamination

The level of contamination of the feedstock material is a key consideration. If the feedstock to the digesters has significant levels of physical contaminants, such as plastic, glass, or metals, then processing to remove the contaminants will be required for the material to be used. If it is not removed, then the digesters can be blocked and will not function efficiently. It is with this understanding that mechanical biological treatment plants are designed. The higher the level of pretreatment a feedstock requires, the more processing machinery will be required, and, hence, the project will have higher capital costs.

After sorting or screening to remove any physical contaminants from the feedstock, the material is often shredded, minced, and mechanically or hydraulically pulped to increase the surface area available to microbes in the digesters and, hence, increase the speed of digestion. The maceration of solids can be achieved by using a chopper pump to transfer the feedstock material into the airtight digester, where anaerobic treatment takes place.

Substrate Composition

Substrate composition is a major factor in determining the methane yield and methane production rates from the digestion of biomass. Techniques to determine the compositional characteristics of the feedstock are available, while parameters such as solids, elemental, and organic analyses are important for digester design and operation.

Applications

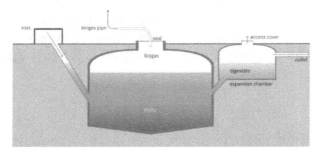

Schematic of an anaerobic digester as part of a sanitation system. It produces a digested slurry (digestate) that can be used as a fertilizer, and biogas that can be used for energy.

Using anaerobic digestion technologies can help to reduce the emission of greenhouse gases in a number of key ways:

- Replacement of fossil fuels
- Reducing or eliminating the energy footprint of waste treatment plants
- Reducing methane emission from landfills
- Displacing industrially produced chemical fertilizers
- Reducing vehicle movements
- Reducing electrical grid transportation losses
- Reducing usage of LP Gas for cooking

Waste and Wastewater Treatment

Anaerobic digestion is particularly suited to organic material, and is commonly used for industrial effluent, wastewater and sewage sludge treatment. Anaerobic digestion, a simple process, can greatly reduce the amount of organic matter which might otherwise be destined to be dumped at sea, dumped in landfills, or burnt in incinerators.

Pressure from environmentally related legislation on solid waste disposal methods in developed countries has increased the application of anaerobic digestion as a process for reducing waste volumes and generating useful byproducts. It may either be used to process the source-separated fraction of municipal waste or alternatively combined with mechanical sorting systems, to process residual mixed municipal waste. These facilities are called mechanical biological treatment plants.

If the putrescible waste processed in anaerobic digesters were disposed of in a landfill, it would break down naturally and often anaerobically. In this case, the gas will eventually escape into the atmosphere. As methane is about 20 times more potent as a greenhouse gas than carbon dioxide, this has significant negative environmental effects.

In countries that collect household waste, the use of local anaerobic digestion facilities can help to reduce the amount of waste that requires transportation to centralized landfill sites or incineration facilities. This reduced burden on transportation reduces carbon emissions from the collection vehicles. If localized anaerobic digestion facilities are embedded within an electrical distribution network, they can help reduce the electrical losses associated with transporting electricity over a national grid.

Power Generation

In developing countries, simple home and farm-based anaerobic digestion systems offer the potential for low-cost energy for cooking and lighting. From 1975, China and India have both had large, government-backed schemes for adaptation of small biogas plants for use in the household for cooking and lighting. At present, projects for anaerobic digestion in the developing world can gain financial support through the United Nations Clean Development Mechanism if they are able to show they provide reduced carbon emissions.

Methane and power produced in anaerobic digestion facilities can be used to replace energy derived from fossil fuels, and hence reduce emissions of greenhouse gases, because the carbon in biodegradable material is part of a carbon cycle. The carbon released into the atmosphere from the combustion of biogas has been removed by plants for them to grow in the recent past, usually within the last decade, but more typically within the last growing season. If the plants are regrown, taking the carbon out of the atmosphere once more, the system will be carbon neutral. In contrast, carbon in fossil fuels has been sequestered in the earth for many millions of years, the combustion of which increases the overall levels of carbon dioxide in the atmosphere.

Biogas from sewage works is sometimes used to run a gas engine to produce electrical power, some or all of which can be used to run the sewage works. Some waste heat from the engine is then used to heat the digester. The waste heat is, in general, enough to heat the digester to the required temperatures. The power potential from sewage works is limited – in the UK, there are about 80 MW total of such generation, with the potential to increase to 150 MW, which is insignificant compared to the average power demand in the UK of about 35,000 MW. The scope for biogas generation from nonsewage waste biological matter – energy crops, food waste, abattoir waste, etc. - is much higher, estimated to be capable of about 3,000 MW. Farm biogas plants using animal waste and energy crops are expected to contribute to reducing CO_2 emissions and strengthen the grid, while providing UK farmers with additional revenues.

Some countries offer incentives in the form of, for example, feed-in tariffs for feeding electricity onto the power grid to subsidize green energy production.

In Oakland, California at the East Bay Municipal Utility District's main wastewater treatment plant (EBMUD), food waste is currently codigested with primary and secondary municipal wastewater solids and other high-strength wastes. Compared to municipal wastewater solids digestion alone, food waste codigestion has many benefits. Anaerobic digestion of food waste pulp from the EBMUD food waste process provides a higher normalized energy benefit, compared to municipal wastewater solids: 730 to 1,300 kWh per dry ton of food waste applied compared to 560 to 940 kWh per dry ton of municipal wastewater solids applied.

Grid Injection

Biogas grid-injection is the injection of biogas into the natural gas grid. The raw biogas has to be previously upgraded to biomethane. This upgrading implies the removal of contaminants such as hydrogen sulphide or siloxanes, as well as the carbon dioxide. Several technologies are available for this purpose, being the most widely implemented the pressure swing adsorption (PSA), water or amine scrubbing (absorption processes) and, in the last years, membrane separation. As an alternative, the electricity and the heat can be used for on-site generation, resulting in a reduction of losses in the transportation of energy. Typical energy losses in natural gas transmission systems range from 1–2%, whereas the current energy losses on a large electrical system range from 5–8%.

In October 2010, Didcot Sewage Works became the first in the UK to produce biomethane gas supplied to the national grid, for use in up to 200 homes in Oxfordshire. By 2017, UK electricity firm Ecotricity plan to have digester fed by locally sourced grass fueling 6000 homes

Vehicle Fuel

After upgrading with the above-mentioned technologies, the biogas (transformed into biomethane) can be used as vehicle fuel in adapted vehicles. This use is very extensive in Sweden, where over 38,600 gas vehicles exist, and 60% of the vehicle gas is biomethane generated in anaerobic digestion plants.

Fertiliser and Soil Conditioner

The solid, fibrous component of the digested material can be used as a soil conditioner to increase the organic content of soils. Digester liquor can be used as a fertiliser to supply vital nutrients to soils instead of chemical fertilisers that require large amounts of energy to produce and transport. The use of manufactured fertilisers is, therefore, more carbon-intensive than the use of anaerobic digester liquor fertiliser. In countries such as Spain, where many soils are organically depleted, the markets for the digested solids can be equally as important as the biogas.

Cooking Gas

By using a bio-digester, which produces the bacteria required for decomposing, cooking gas is generated. The organic garbage like fallen leaves, kitchen waste, food waste etc. are fed into a crusher unit, where the mixture is conflated with a small amount of water. The mixture is then fed into the bio-digester, where the bacteria decomposes it to produce cooking gas. This gas is piped to kitchen stove. A 2 cubic meter bio-digester can produce 2 cubic meter of cooking gas. This is equivalent to 1 kg of LPG. The notable advantage of using a bio-digester is the sludge which is a rich organic manure.

Products

The three principal products of anaerobic digestion are biogas, digestate, and water.

Biogas

Typical composition of biogas		
Compound	Formula	%
Methane	CH_4	50–75
Carbon dioxide	CO_2	25–50
Nitrogen	N_2	0–10
Hydrogen	H_2	0–1
Hydrogen sulphide	H_2S	0–3
Oxygen	O_2	0–0
Source: *www.kolumbus.fi, 2007*		

Biogas is the ultimate waste product of the bacteria feeding off the input biodegradable feedstock (the methanogenesis stage of anaerobic digestion is performed by archaea (a micro-organism on a distinctly different branch of the phylogenetic tree of life to bacteria), and is mostly methane and carbon dioxide, with a small amount hydrogen and trace hydrogen sulfide. (As-produced, biogas also contains water vapor, with the fractional water vapor volume a function of biogas temperature). Most of the biogas is produced during the middle of the digestion, after the bacterial population has grown, and tapers off as the putrescible material is exhausted. The gas is normally stored on top of the digester in an inflatable gas bubble or extracted and stored next to the facility in a gas holder.

The methane in biogas can be burned to produce both heat and electricity, usually with a reciprocating engine or microturbine often in a cogeneration arrangement where the electricity and waste heat generated are used to warm the digesters or to heat buildings. Excess electricity can be sold to suppliers or put into the local grid. Electricity produced by anaerobic digesters is considered to be renewable energy and may attract subsidies. Biogas does not contribute to increasing atmospheric carbon dioxide concentrations because the gas is not released directly into the atmosphere and the carbon dioxide comes from an organic source with a short carbon cycle.

Biogas may require treatment or 'scrubbing' to refine it for use as a fuel. Hydrogen sulfide, a toxic product formed from sulfates in the feedstock, is released as a trace component of the biogas. National environmental enforcement agencies, such as the U.S. Environmental Protection Agency or the English and Welsh Environment Agency, put strict limits on the levels of gases containing hydrogen sulfide, and, if the levels of hydrogen sulfide in the gas are high, gas scrubbing and cleaning equipment (such as amine gas treating) will be needed to process the biogas to within regionally accepted levels. Alternatively, the addition of ferrous chloride $FeCl_2$ to the digestion tanks inhibits hydrogen sulfide production.

Volatile siloxanes can also contaminate the biogas; such compounds are frequently found in household waste and wastewater. In digestion facilities accepting these materials as a component of the feedstock, low-molecular-weight siloxanes volatilise into biogas. When this gas is combusted in a gas engine, turbine, or boiler, siloxanes are converted into silicon dioxide (SiO_2), which deposits internally in the machine, increasing wear and tear. Practical and cost-effective technologies to remove siloxanes and other biogas contaminants are available at the present time. In certain applications, *in situ* treatment can be used to increase the methane purity by reducing the offgas carbon dioxide content, purging the majority of it in a secondary reactor.

In countries such as Switzerland, Germany, and Sweden, the methane in the biogas may be compressed for it to be used as a vehicle transportation fuel or input directly into the gas mains. In countries where the driver for the use of anaerobic digestion are renewable electricity subsidies, this route of treatment is less likely, as energy is required in this processing stage and reduces the overall levels available to sell.

Biogas holder with lightning protection rods and backup gas flare

Biogas carrying pipes

Digestate

Digestate is the solid remnants of the original input material to the digesters that the microbes cannot use. It also consists of the mineralised remains of the dead bacteria from within the digesters. Digestate can come in three forms: fibrous, liquor, or a sludge-based combination of the two fractions. In two-stage systems, different forms of digestate come from different digestion tanks.

In single-stage digestion systems, the two fractions will be combined and, if desired, separated by further processing.

Acidogenic anaerobic digestate

The second byproduct (acidogenic digestate) is a stable, organic material consisting largely of lignin and cellulose, but also of a variety of mineral components in a matrix of dead bacterial cells; some plastic may be present. The material resembles domestic compost and can be used as such or to make low-grade building products, such as fibreboard. The solid digestate can also be used as feedstock for ethanol production.

The third byproduct is a liquid (methanogenic digestate) rich in nutrients, which can be used as a fertiliser, depending on the quality of the material being digested. Levels of potentially toxic elements (PTEs) should be chemically assessed. This will depend upon the quality of the original feedstock. In the case of most clean and source-separated biodegradable waste streams, the levels of PTEs will be low. In the case of wastes originating from industry, the levels of PTEs may be higher and will need to be taken into consideration when determining a suitable end use for the material.

Digestate typically contains elements, such as lignin, that cannot be broken down by the anaerobic microorganisms. Also, the digestate may contain ammonia that is phytotoxic, and may hamper the growth of plants if it is used as a soil-improving material. For these two reasons, a maturation or composting stage may be employed after digestion. Lignin and other materials are available for degradation by aerobic microorganisms, such as fungi, helping reduce the overall volume of the material for transport. During this maturation, the ammonia will be oxidized into nitrates, improving the fertility of the material and making it more suitable as a soil improver. Large composting stages are typically used by dry anaerobic digestion technologies.

Wastewater

The final output from anaerobic digestion systems is water, which originates both from the moisture content of the original waste that was treated and water produced during the microbial reactions in the digestion systems. This water may be released from the dewatering of the digestate or may be implicitly separate from the digestate.

The wastewater exiting the anaerobic digestion facility will typically have elevated levels of biochemical oxygen demand (BOD) and chemical oxygen demand (COD). These measures of the reactivity of the effluent indicate an ability to pollute. Some of this material is termed 'hard COD', meaning it cannot be accessed by the anaerobic bacteria for conversion into biogas. If this effluent were put directly into watercourses, it would negatively affect them by causing eutrophication. As such, further treatment of the wastewater is often required. This treatment will typically be an oxidation stage wherein air is passed through the water in a sequencing batch reactors or reverse osmosis unit.

History

Gas street lamp

Reported scientific interest in the manufacturing of gas produced by the natural decomposition of organic matter dates from the 17th century, when Robert Boyle (1627-1691) and Stephen Hales (1677-1761) noted that disturbing the sediment of streams and lakes released flammable gas. In 1808 Sir Humphry Davy proved the presence of methane in the gases produced by cattle manure. In 1859 a leper colony in Bombay in India built the first anaerobic digester. In 1895, the technology was developed in Exeter, England, where a septic tank was used to generate gas for the sewer gas destructor lamp, a type of gas lighting. Also in England, in 1904, the first dual-purpose tank for both sedimentation and sludge treatment was installed in Hampton, London. In 1907, in Germany, a patent was issued for the Imhoff tank, an early form of digester.

Research on anaerobic digestion began in earnest in the 1930s.

References

- Zehnder, Alexander J. B. (1978). "Ecology of methane formation". In Mitchell, Ralph. Water pollution microbiology 2. New York: Wiley. pp. 349–376. ISBN 978-0-471-01902-2.

- "COMPARING OF MESOPHILIC AND THERMOPHILIC ANAEROBIC FERMENTED SEWAGE SLUDGE BASED ON CHEMICAL AND BIOCHEMICAL TESTS" (PDF). www.aloki.hu. Retrieved 23 February 2016.

- Dosta, Joan; Galí, Alexandre; Macé, Sandra; Mata-Álvarez, Joan (February 2007). "Modelling a sequencing batch reactor to treat the supernatant from anaerobic digestion of the organic fraction of municipal solid waste". Journal of Chemical Technology & Biotechnology. 82 (2): 158–64. doi:10.1002/jctb.1645. Retrieved 16 September 2013.

- Gupta, Sujata (2010-11-06). "Biogas comes in from the cold". New Scientist. London: Sunita Harrington. p. 14. Retrieved 2011-02-04.

- Tower, P.; Wetzel, J.; Lombard, X. (March 2006). "New Landfill Gas Treatment Technology Dramatically Lowers Energy Production Costs". Applied Filter Technology. Archived from the original on 24 September 2011. Retrieved 2009-04-30.

- National Non-Food Crops Centre. "NNFCC Renewable Fuels and Energy Factsheet: Anaerobic Digestion", Retrieved on 2011-11-22

- Shah, Dhruti (5 October 2010). "Oxfordshire town sees human waste used to heat homes". BBC NEWS. Archived from the original on 5 October 2010. Retrieved 5 October 2010.

Permissions

All chapters in this book are published with permission under the Creative Commons Attribution Share Alike License or equivalent. Every chapter published in this book has been scrutinized by our experts. Their significance has been extensively debated. The topics covered herein carry significant information for a comprehensive understanding. They may even be implemented as practical applications or may be referred to as a beginning point for further studies.

We would like to thank the editorial team for lending their expertise to make the book truly unique. They have played a crucial role in the development of this book. Without their invaluable contributions this book wouldn't have been possible. They have made vital efforts to compile up to date information on the varied aspects of this subject to make this book a valuable addition to the collection of many professionals and students.

This book was conceptualized with the vision of imparting up-to-date and integrated information in this field. To ensure the same, a matchless editorial board was set up. Every individual on the board went through rigorous rounds of assessment to prove their worth. After which they invested a large part of their time researching and compiling the most relevant data for our readers.

The editorial board has been involved in producing this book since its inception. They have spent rigorous hours researching and exploring the diverse topics which have resulted in the successful publishing of this book. They have passed on their knowledge of decades through this book. To expedite this challenging task, the publisher supported the team at every step. A small team of assistant editors was also appointed to further simplify the editing procedure and attain best results for the readers.

Apart from the editorial board, the designing team has also invested a significant amount of their time in understanding the subject and creating the most relevant covers. They scrutinized every image to scout for the most suitable representation of the subject and create an appropriate cover for the book.

The publishing team has been an ardent support to the editorial, designing and production team. Their endless efforts to recruit the best for this project, has resulted in the accomplishment of this book. They are a veteran in the field of academics and their pool of knowledge is as vast as their experience in printing. Their expertise and guidance has proved useful at every step. Their uncompromising quality standards have made this book an exceptional effort. Their encouragement from time to time has been an inspiration for everyone.

The publisher and the editorial board hope that this book will prove to be a valuable piece of knowledge for students, practitioners and scholars across the globe.

Index

www.ingramcontent.com/pod-product-compliance
Lightning Source LLC
Jackson TN
JSHW052121213015
770331S0004B/233